GEOGRAPHY OF INDIA

GEOGRAPHY OF INDIA

COMPRISING AN ACCOUNT OF BRITISH INDIA AND THE VARIOUS STATES ENCLOSED OR ADJOINING

ANONYMOUS

GYAN

Published by

Gyan Publishing House
5, Ansari Road
Daryaganj, New Delhi-110002
Phone: 011-47034999, 9811692060
E-mail: books@gyanbooks.com

Distribution Network
gyanbooks.com
India, USA, Canada, UK, Australia

ISBN: 978-93-7169-685-2 (PB)

First Published, 1870

2nd Impression 2025

Printed at: Gyan Press, Delhi.

GEOGRAPHY OF INDIA
Author: ANONYMOUS

GEOGRAPHY OF INDIA;

COMPRISING AN ACCOUNT OF

BRITISH INDIA,

AND

THE VARIOUS STATES ENCLOSED OR ADJOINING.

FOR THE USE OF SCHOOLS AND STUDENTS.

LONDON:
WM. H. ALLEN & CO., 13, WATERLOO PLACE, S.W.
1870.

INTRODUCTORY NOTICE.

In the absence of any convenient and correct account of Modern Indian Geography, the following pages are submitted with some confidence as adapted for general use both in England and India.

The admirable gazetteers of Mr. Thornton, based on information obtained from the most accurate sources, and prepared with great care, have served as the basis of the present work, and have been freely drawn upon. Various blue books, and statements published by authority, some of them during the present year, have been made use of to bring the accounts down to a recent period, and changes in the distribution of districts and provinces are duly noted. Other works on India have been referred to, and valuable assistance has been obtained, in the revision of the pages, from those best informed on the subject in this country.

The spelling of proper names has been rendered as simple as possible. Unnecessary letters have been suppressed. Thus *u* replaces *oo*, and *i* replaces *ee*; *h* is generally omitted after *d* and *g*, and at the ends of words. Both *h* and *r* are left out where the pronunciation is not affected.

A full Index, containing more than 2,000 references, and a careful Table of Contents, will render the work useful for reference. It is hoped that it may be found available as a class-book, not only for English instruction but in native schools where English is taught.

D. T. A.

CONTENTS.

CHAPTER I.

General Description.

CHAPTER II.

Northern India.—Eastern Division.

CHAPTER III.

Northern India.—Western Division.

CHAPTER IV.

Central India.

CHAPTER V.

Western India.

CHAPTER VI.

Southern India.

GEOGRAPHY OF INDIA.

CHAPTER I.

GENERAL DESCRIPTION.

Position, Boundaries, and Extent—Mountains—Plateaux—Plains and Valleys—Rivers—Lakes and Lagoons—Climate—Vegetable and Animal Life—Mineral Resources—Population and Races of Men—Language—Religions—Government—Revenue—Trade and Commerce—Means of Communication—Canals and Irrigation Works—Political and Natural Divisions—Progress of British Power in India.

Position, Boundaries, and Extent.—The peninsula of India occupies nearly the same relative position with regard to Asia that Italy does to Europe. It consists of a spur projecting southwards from the central and loftiest part of the great mountain axis of the Old World. This spur is wide towards the mountains, and gradually narrows till it terminates southward at Cape Comorin. The large island of Ceylon represents Sicily. The great River Ganges conveys the drainage of the southern face of the Asiatic Alps into the Bay of Bengal, just as the Po does that of the European Alps into the Adriatic. The mountains of Arracan on the east, and of the country beyond the Indus to the west, form, with the Himalaya, the land boundaries of India; and within these limits and the shores of the Indian Ocean and the Bay of Bengal is contained the whole country, reaching from about the 36th to the 8th parallel of north latitude. Its limits of longitude are 66½° and 99½° east from Greenwich. Unlike Italy, however, in some respects, much of the central part of India is table-land; mountains rising chiefly on the western, but partly on the eastern shore, and occasionally crossing the country.

Both Italy and India consist largely of volcanic rock; but in India itself are no active volcanoes. The climate of India, affected by its geographical position, is necessarily very different from that of any part of Europe; and the result of this is seen in totally distinct natural groups of plants and animals, and even in the races of men who inhabit the great Asiatic promontory.

India proper, or *Hindustan*, is separated from Tibet to the north by the Himalayan mountains; from China and the Burman Empire to the north-east by spurs of the Eastern Himalaya, serving as the eastern watershed of the Bramaputra; and from Affganistan and Beluchistan to the north-west by spurs of the Western Himalaya (the *Suliman* range), serving as the western watershed of the Indus. Naturally and geographically, the peninsula commences on the east at the mouths of the Ganges, and on the west at the mouth of the Indus; but, politically, India includes Arracan, Pegu, and the Tenasserim provinces, the Yoma mountains being the boundary of the former, while the Siamese mountains separate Tenasserim from Siam. The Arabian Sea washes the western, and the Bay of Bengal the eastern shores of the peninsula, and its southern extremity projects into the Indian Ocean. Its greatest length (not including Ceylon), measured from the extremity of the Punjab on the north, to Cape Comorin in the south, is 1,830 miles. Its breadth from Kurrachi in the west, beyond the Indus, to the eastern extremity of Assam, beyond the Bramaputra, is nearly as much. Its total area is estimated at 1,572,386 square miles. From the Gulf of Cambay to the mouth of the Hoogly—which may be regarded as the limits of the smaller and more distinct peninsula—the distance is nearly 1,000 miles.

India has a coast-line of about 3,600 miles. The south-west coast is called the Malabar coast, from the province of that name. The east coast from Palk's Straits to the mouth of the Kistna, is called the Coromandel coast, from the name of an ancient Indian kingdom. From the mouth of the Kistna to the mouths of the Godavery it is called the Golconda coast. Beyond that to the Hoogly is the Orissa coast.

Notwithstanding the great length of coast-line round India, the number of its harbours is few, and the ports, though numerous, are not well adapted for trading purposes: Bombay; Coringa, on the Coromandel coast; Cochin, on the Malabar coast; and Trincomali, in Ceylon, are the best natural harbours. The chief ports are Aleppi, Beypur Bimlipatam, Bombay, Calicut, Cannanore, Cochin, Coconada, Coringa, Karwa, Kurrachi, Madras, Mangalore Masulipatam, Moulmein, Negapatam, Pondicherry, Quilon Rangoon, Tellicherry, Tuticorin, and Vizagapatam.

India includes, in addition to the British possessions, nearly two hundred native states, varying, in dimensions, from a few hundred acres to nearly 100,000 square miles dispersed sometimes singly, sometimes in groups, and generally surrounded by, or included in, British districts. These are governed by native chiefs,* under the protection of the British Government, and almost all pay tribute. The total area of these states (including French and Portuguese possessions) is nearly 670,000 square miles, leaving 904,628 square miles of British possessions. Of the native states, a part (about 50,000 square miles), is under British administration.

Mountains.—1. The *Himalaya chain* is the great mountain chain of India, and separates the peninsula from Central Asia. It stretches, in an irregular line,

---

* The Hindu chiefs are called Rajas and Maharajas; the Mohamedan, Nawabs.

from the defile above Cashmere on the north-west (through which the Indus penetrates to the plains of the Punjab, dividing the chain from that of the Hindu Cush) to the southern bend by which the Sanpu or Dihong river enters India to join the Bramaputra. The Indus and Bramaputra start from almost the same point on the northern side of the Himalaya, the one proceeding west, and the other east; these river valleys nearly detach the chain. It extends over 22 degrees of longitude. Between the sources of the Indus and Bramaputra is a great mountain knot, which connects the chain with others to the north. The mean breadth of the Himalaya is about 150 miles, and the mean height is estimated at 20,000 feet. The height of the snow-line on the Indian side is 18,500 feet. The greatest elevation in the Himalaya is Mount Everest, 29,000 feet above the sea; near it are several peaks—Kinchinjunga and Dawalagiri, both upwards of 28,000 feet, and a large number above 25,000. The passes across the Himalaya, from the valleys on the Indian side, are all lofty and difficult of access, 17,000 feet is about the usual height.

In front of the main Himalayan chain are subordinate or sub-Himalayan ranges, and between the last of these and the plains of India is a broad strip of marshy land, covered with forest and jungle, called the "*Terai*" (*see* page 100). It is subject to the most fatal malaria to human beings, but is crowded with wild animals of all kinds. It also contains much rich soil and some iron ore.

2. *Sewalik range.*—These mountains are parallel to the Himalaya between the Ganges and the Sutlej for about 155 miles. Their breadth is about ten miles, and their highest point is 3,500 feet above the sea. Towards the Sutlej they are low hills. They are remarkable for the interesting fossils that have been found in them, indicating the nature of the former inhabitants of this part of the world.

3. *Suliman Mountains.*—This extensive and lofty range

forms the boundary of British empire in Western India. It extends from south to north 350 miles (from lat 29° to 33° 50′), and is nearly parallel to the course of the Indus, from which it is distant seventy miles. The extreme elevation is about 11,000 feet, which is below the limit of perpetual snow in that latitude. The steepest, or escarped side, is towards the Indus, the western declivity being gradual, and passing into the desert table-land of Sivestan, in Affganistan. Many streams descend the eastern side, and fertilize the plains at their foot, but hardly any of the water reaches the Indus, and no streams whatever proceed from the other side. Vegetation on these mountains is vigorous and varied, the sides being covered with lofty and dense forest, and the valleys overgrown with flowering plants.

4. *Salt Range.*—A group of mountains ranging south-eastwards from the northern extremity of the Suliman mountains to the right bank of the River Jelum. It receives its English name from its numerous and thick beds of common salt, which have been long worked to great advantage. The elevation is not considerable, nowhere exceeding 3,000 feet above the sea. The range is crossed by the Indus, which has cut a deep rocky channel, on the sides of which the salt-beds are exposed. Besides salt, this range contains much iron ore of the richest kind, and there is a considerable quantity of gold. Alum, slate, sulphur, saltpetre, and other earthy minerals have been observed (*see* page 122).

5. *Vindya Range.*—This is also an east and west range, and it separates the valley of the Nerbudda, and the other parts of Central India, from the valleys of the Jumna and Ganges. It unites, also, the northern extremities of the Western and Eastern Ghats. The range is nearly 150 miles in length, and it consists of granite and sandstone, capped by basalt. The mountains descend on the northern side into the table-land of Malwa, and may be traced eastwards—though of no great elevation—crossing the valleys of the Ganges and Brama-

putra, and rising again in the mountains of Assam. Much of the country is rugged, but the elevations are not very considerable. The summits vary in height from 1,000 to 2,000 feet, the crest of the Jam Ghat, the highest point, being 2,328 feet above the sea.

6. *Satpura Mountains.*—An east and west range of about 200 miles, parallel to the Vindya range, dividing the valley of the Nerbudda from that of the Tapti. The south declivity, towards the Tapti, is abrupt; the northern, towards the Nerbudda, very gentle. The appearance of the chain is more bold and romantic than that of the Vindya, rising into lofty peaks, and swelling into fine forms. They are of volcanic origin, consisting of greenstone and other modern igneous rock, capped with basalt. Asirgur, one of the summits, is about 1,200 feet above the sea. The highest summits are estimated at 2,000 feet.

7. *Mahadeo Mountains.*—A cluster of mountain land, of considerable height, in the northern part of the Nagpur territory, situated at the eastern extremity of the Satpura mountains, where they adjoin the Vindya, and extending from long. 78° to 80°. The higher elevations have been estimated at 5,000 feet, but are probably much less. The culminating ridge forms the watershed separating the waters that go to the Nerbudda from those of the Godavery.

8. *Aravalli Mountains.*—A long low chain, running north-east, between the basins of the Ganges and the Indus, and touching the western extremity of the Vindya. It is bold and precipitous on the north-west side, but less so on the south-east. It is without any pass for wheel-carriages for 220 miles. The highest summit is Mount Abu, estimated at 5,000 feet above the sea; but the usual elevation is about 3,600 feet.

9. *Western Ghats.*—The name Ghat, or Ghaut,* is

---

* The word *Ghat* properly designates passes across mountain ·ins rising in successive steps.

given in India to the mountain chains that ascend rapidly with a very rough and bold escarpment from the coast, chiefly on the western, but also on the eastern side of the peninsula, south of the Tapti on the one side, and the Bengal provinces on the other. The Western is by far the loftiest and the most important chain, commencing with elevations of about 2,000 feet in lat. 21° 10′, and proceeding southward, gradually becoming more and more lofty, for more than 200 miles, when its height, at Mahabaleshwar, is 4,700 feet. Here, as elsewhere, the western declivity is abrupt, and its base depressed nearly to the sea-level, a distance rarely exceeding forty or fifty miles from the ridge. On the eastern side of the watershed the range descends to a lofty rugged table-land of considerable elevation, gradually, but very slowly, depressed towards the plains of Hyderabad. To the south of Mahabaleshwar, the height of the range diminishes for about 200 miles, until in lat. 15°, on the frontier of the Madras Presidency, it is not more than 1,000 feet, the slopes being gradual and the outlines rounded. Still further south the elevation again increases, and attains its maximum near Coorg, where Bonasson Hill is about 7,000 feet, and there are other summits of nearly 6,000. The Ghats at this point join and pass into the Neilgherry mountains, and afterwards rise into the *Kunda Mountains*, which terminate abruptly to the south in high and almost vertical precipices, extending across the peninsula, and connecting the Western with the Eastern Ghats. There is a gap twenty miles wide between this steep escarpment and the continuation of the range southward, which terminates in Cape Comorin, whose summits attain a height of from 5,000 to 7,000 feet.

The lofty range thus extending for about 800 miles near the coast is broken at intervals by gorges and chasms, which give access to the highlands and the plains of Central and Southern India, and which give their name to the whole chain. These passes are not very lofty. The whole range of the Western Ghats present no summits which can compare with even the lower ranges of the Himalaya chain, but the character of the scenery is exceedingly grand, and many of the higher peaks are almost inaccessible, owing to the extreme ruggedness of the outline. The rocks are to a very large extent volcanic; enormous sheets of basalt ranging for very long distances. No streams break through the Western Ghats, the only drainage being that from the western slope of the chain.

10. *Eastern Ghats.*—This chain is both lower and less abrupt than the Western, and differs from it in being more distant from the coast, rising with less abruptness from the sea, and giving access, by a number of deep gorges and wide openings, to the drainage of the whole of the interior peninsula. The average elevation is estimated at 1,500 feet, the highest summit (near Madras) being about 3,000 feet. The Eastern and Western Ghats connect in the south, as already stated. The main foundation of the Eastern Ghats is granitic rock.

11. *Neilgherries.*—The small mountain chain thus named is really a lofty spur at the southern extremity of the Western Ghats. The range extends towards the north-east for about forty miles, the extreme breadth being fifteen miles. The northern side, towards Mysore, rises about 3,500 feet above that table-land, and connects, by a series of precipitous granite peaks, with the Ghats. There are three parts of the range, the central rising, in the Mountain of Dodabetta, to 8,760 feet, the culminating point of all the mountain land of Southern India. There are five other peaks above 8,000 feet, and ten between 6,000 and 8,000 feet (*see* note, p, 214).

Plateaux.—The central part of India consists of the lofty plateau of the Deccan, extending between the Eastern and Western Ghats, and gradually sloping to the south-east. Much of it is covered by basalt, the parts thus capped being generally preserved from the denuding action of the weather, and producing the isolated flat-topped hills so characteristic of the country. The *Pulnai Hills* are a lofty part of this plateau in the south, rising in parts to 8,000 feet, and generally 7,000 feet, above the sea. They are very picturesque. The mean elevation of the Deccan is about 3,000 feet, from which it sinks to 1,500 feet. The central part is composed of undulating plains, without trees, covered with verdure after rain, but desert in the dry season. North of the Nerbudda there is another plateau, that of Malwa. The plateau of Malwa is from 1,500 to 2,500 feet above the sea.

Plains and Valleys.—A very large part of Northern India consists of the valleys of the Ganges and Bramaputra, extending between the lowest ranges of the Himalaya to the north, the table-land of Malwa and its extension eastward to the south, and the Aravalli mountains to the west. This vast space rises gradually from the east towards the west and north-west; but for the most part does not attain an elevation of more then 1,000 feet, except where it approaches the mountains. There is also an extensive but narrow plain between the Eastern Ghats and the sea, on the coast of the Carnatic, and the tongue of land between the Kistna and Godavery (also low plain) is of considerable extent. Between the Aravalli mountains and the Suliman mountains is the valley of the Indus, which includes a large tract of sandy desert called the *Thur*, with no eminence except hills of blown sand. This district extends between the left bank of the Indus and the mountains, and includes a large part of Rajputana, and some part of Sind. The Runns of Cutch are included in this area (*see* p. 194).

Rivers.—The drainage system proceeding from the Himalaya mountains, corresponds in magnitude and importance to the vast rocky mass from whose valleys and glaciers the water is derived. It comprises three of the principal rivers of the Western hemisphere, with their numerous tributary streams; and each stream affords a complete network of waters.

1. The INDUS.—This stream takes its rise in lat. 32°, long. 81° 30′, a little north of the Kailas mountain, at an elevation of 22,000 feet above the sea. It flows over a table-land for about 100 miles, receiving several tributaries, and then enters the deep gorges of a great depression, separating the Kuen Lun mountains from the Himalaya chain. It first receives the name of Indus in Little Tibet, where the *Shayuk* joins it at a distance of about 500 miles from the source. As far as that point the name is *Sin-ka-bab*, or the Lion's Mouth. At the junction the Shayuk is the larger river, but the other the deeper and fuller stream. Not far below this junction the Indus emerges from the mountain region, passing the western side of Cashmere. It begins to be navigable at Attock, about 870 miles from the source, and 940 from the sea. At this point it receives the *Cabul* river, a considerable stream, from the west. The Indus is here 800 feet wide, sixty feet deep, and runs at the rate of six miles an hour at some seasons, and is about 1,000 feet above the sea. Between Attock and Kala Bag (less than 100 miles), it runs between precipices and in ravines, and then enters the plain. From this point it loses its clearness, and is subject to seasonal inundations, commencing with the spring, and caused by the melting of snow. It continues now with a steady southern course, and with little addition for about 300 miles, till its confluence with the channel which conveys the Sutlej and the other large streams of the Punjab. At this confluence the affluent is the wider stream, but the Indus conveys the greater body of water. Below the junction of the Punjab group of rivers the Indus is never less than 2,000 yards wide, but it is not very deep. It passes for about 350 miles through a plain country, shifting its course, and often expanding, with numerous branches, till it reaches Tatta, the head of the existing delta. The bulk of the water is discharged by a few large but shifting courses along a line of about 130 miles of coast, at the head of the Arabian Sea. The *Kukywari* and the *Kori* are the widest channels; but inland navigation is kept up by

a multitude of ways, varying from one season to another. When the river is high, the whole of the mouths communicate more or less freely, and are entirely under water. The tide reaches Tatta, about seventy miles from the sea, and the spring tide rises nine feet. The whole course of the stream is about 1,800 miles. The drainage area of the Indus is 415,000 square miles.

The rivers of the Punjab, combining before they reach the Indus, are the only important affluents of that river in the lower part of its course, and are delivered in a group of channels forming a kind of inland delta. The main streams, commencing with that most to the east, are the Sutlej, the Beas, the Ravi, the Chenab, and the Jelum. The *Sutlej* takes its rise in the interior of the Himalaya, in lat. 30° 8′, long. 81° 53′, a little north of Nepaul, and not far from the source of the Indus. It first collects in the sacred lakes of Manasarovara and Rawan Hrad, and after its emergence from these, soon becomes an important stream. It only receives its name of Sutlej at some distance from this point, and after having received several feeders. At first it is a raging torrent, falling in some places 150 feet in a mile. It receives the *Li* or *Spiti* river, a principal tribute, and emerges to the wild mountain country at a level of about 8,000 feet above the sea. It then runs, at first south-west, and afterwards west, with a very great declivity, passing between the sources of the Beas, one of its principal confluents, and of the great River Jumna. It then continues, without large contributions, passing through several of the small states known as the Hill States, until it enters the plains of the Punjab at Rupur. Beyond this point it is navigable at all seasons. About 100 miles lower down it is joined by the *Beas*, whose stream is wider and larger than that of the Sutlej, but whose course is entirely through the lower ranges of the mountains, and is much shorter. The length of the Sutlej, from its source to the Beas, is 550 miles. Below this junction the river runs 300 miles, under the name of *Gara*, to its confluence with the combined streams of the Chenab and the Jelum. The Ravi enters the Chenab below the confluence of these two streams, but before they enter the Sutlej. The *Ravi* takes its rise in the Hill States, a little west of the Beas, and has an exceedingly tortuous course through the plains of the Punjab, which it traverses through precipitous banks, and is deep, but not wide. From Lahore, which it passes, it continues 200 miles in a direct line (but 380, including windings) to the Chenab, which it enters by three mouths. Its total length is about 450 miles, without including windings.

The *Chenab* is the largest of the five rivers that traverse the Punjab. It rises, in Cashmere, 13,000 feet above the sea, and receives several tributaries before entering the Punjab. It is navigable for timber rafts before leaving Cashmere. Down to its junction with the Jelum it has a course of 605 miles, and is a vast and important stream at all seasons. The *Jelum* or *Behut*, the most western of the great rivers of the Punjab, also rises in Cashmere, and drains the valley of Cashmere. It passes through several lakes, and is navigable for a long distance. It leaves the mountains about 250 miles from its source, and enters the plain of the Punjab where it joins the Chenab, after a further course of nearly 200 miles. Below the confluence the Chenab receives the name of *Tromba* till the Sutlej is reached, after which the name given to the combined waters till they enter the Indus is the *Punjnud*.

2. The GANGES.—This river was formerly considered to originate in a glacier of the Eastern Himalaya, at an elevation of about 14,000 feet above the sea, in the British territory of Gurwal, in lat. 30° 54′, long. 79° 7′. The stream near this source is called the *Baghiretti*. Another stream yielding a large volume of water, and called the *Aluknanda*, is, however, now generally regarded as the principal. Both these drain valleys of the Himalaya, a little south of the feeders of the Indus and Sutlej. Descending at first very rapidly, it issues from the mountains at Suki (lat. 30° 59′, long. 78° 45′), at the height of 7,608 feet above the sea, and thence continues with a SE. course, receiving the *Ramgunga* and a few small feeders,

chiefly from the left bank, till at Allahabad, 670 miles from its mountain sources, it receives the Jumna, a tributary almost superior in importance to the Ganges itself, and bringing water from the north-west and west, besides draining the table-land of Malwa in the south-west. From Allahabad to Benares (140 miles) the combined stream of the Ganges and Jumna meanders through vast plains, its volume of water being subject to enormous and rapid change from the influence of floods. In this part of its course its mean discharge has been estimated at 250,000 cubic feet of water per second. Near Benares it receives the Gumti, and a little below the Gagra river, both from the north, and the Sone river from the south. Beyond these the Gunduck and the Kosi, both from the north, are the only considerable streams till the waters of the Ganges mingle with those of the Bramaputra, and both make their way, by innumerable channels, to the sea. The branches of the Ganges, near its mouth, form a complicated system called the *Sunderbunds*, the combined delta of the Ganges and Bramaputra (*see* p. 60). This delta is 180 miles in width from east to west, and 200 miles from north to south. The *Bagarati* and the *Jellinghi* are the main branches that form the Hoogly. The *Mutla* is the principal navigable branch of the Ganges, and on it is the town of Mutla, or Port Canning. The total length of the Ganges is 1,500 miles, and it is navigable for boats for 1,300 miles of its course. A great and rapid tidal wave, or *bore*, rushes in from the sea, where it is sometimes ten feet high, and at Calcutta this wave is sometimes five feet.

The *Jumna* rises in the Hill States, on the south-western side of the mountains that supply water to the upper streams of the Ganges, and not far from the course of the Upper Sutlej. It receives, in its progress through the mountain country, a number of small streams, and soon becomes a great river. It is, however, a torrential stream, not well adapted to navigation; constantly receiving accessions from both sides, and generally shallow and rocky. After a course of 860 miles it unites with the Ganges at Allahabad, bringing nearly the same volume of water. The Jumna water, though clear as crystal, is less palatable and wholesome than that of the comparatively muddy stream, the Ganges.

The *Chumbul* is an important tributary of the Jumna. It rises in Malwa, 2,000 feet above the sea, among a group of summits of the Vindya range. It soon becomes a considerable stream, and receives tributaries of importance. Further down it passes through magnificent scenery; and at length, after expanding into a lake, falls, in a succession of rapids, a total height of about 200 feet. Below the rapids, at the city of Kota, 260 miles from the source, it is a large deep stream, very difficult to cross, and with a large volume of water, but not very navigable. It enters the Jumna after a course of 570 miles. The *Kali-Sind* also rises in Malwa, and has a course of about 260 miles to the Jumna, receiving many small tributaries. The *Betwa* rises in Bopal, and has a course of 360 miles to the Jumna. It has a rocky bed, and is not navigable. It is an enormous river during the rains.

The *Gumti* is a stream, valuable for navigation and irrigation, rising in the district of Shajehanpore at about 520 feet above the sea, running past Lucknow as a navigable stream, and reaching the Ganges, a little below Benares, after a course of nearly 500 miles, chiefly through the territory of Oude. The *Gagra* rises in the British district of Kumaon, near the Tibet frontier, at an elevation between 17,000 and 18,000 feet. It soon receives several torrents, and becomes a considerable stream where it enters the territory of Oude. Up to this point it is called the *Kali* or *Sarju*. It passes the town of Oude, where it is from one to three miles in breadth, sending off very numerous lateral branches. Further down it receives the River Rapti, and finally enters the Ganges a little above Patna, after a course of 600 miles, and with a volume of water bearing a comparison with that stream both in breadth and rapidity, and sometimes in depth. It is not very navigable. The *Sone* rises in the high table-land of Nagpur, near the source of the Nerbudda, and after a course of 465 miles enters the Ganges nearly opposite the mouth of the Gagra. The bed of the stream is in some places two miles wide, and is filled during the periodical rains. The river is not navigable. The *Gunduck* is another important tributary. It comes in from the north, and is a grand river, taking its rise in Tibet, flowing through the wild valleys of the Himalaya, and entering the Ganges near Patna, its course being about 400 miles. It is not only a large stream, but is navigable for a considerable distance. After the Gunduck the *Cosy*, or *Kosy*, is the only large feeder of the Ganges above the delta. It rises in Nepaul, among the snowy peaks of the Himalaya, and soon becomes a torrent, issuing from the mountains as a first-class stream, nowhere fordable in ordinary years. It descends by three cataracts, unlike most of the Indian rivers. The total length of course is only 325 miles, but it is everywhere a large stream. In its lower part it sends off several branches. The *Hoogly* is the only remaining river of importance in the Ganges system. It is chiefly formed by the junction of two great branches of the Ganges; but it also receives tributaries from the west, of which the *Cossye* and the *Damuda* are the largest. The former has a course of 240 miles.

3. The BRAMAPUTRA.—The main branch of this river rises in Tibet, not far from the sources of the Indus and the Sutlej; and, under the name of the *Sanpu* river, has an easterly course of at least 1,600 miles. It then turns south, entering Assam under the name of the *Dihong* river, and receiving the stream which is afterwards regarded as the principal, though its course is much shorter. Passing through Assam, which it intersects, it receives many tributaries, occasionally branching, and giving rise to islands, some of them of very large dimensions. After flowing in this manner westwards for several hundred miles, it makes a circuit round the Garrow Hills, receiving the *Tista* from the north-west, and then becomes dispersed in an infinite complication of branches, of which the *Kirtynassa* is the principal. The

Tista gives off an important branch named the *Attree*, called the *Balasar* near its mouth. What remains of the stream enters the Bay of Bengal by the *Megna*, a channel parallel to that of the Ganges, at a short distance to the east, and forms, with the Ganges, a network of channels through the alluvial matter brought down by all the rivers. The *Barak* is a considerable feeder of the Megna. The various branches afford internal navigation throughout Bengal. The total length of the Bramaputra, measuring only from the eastern and shorter branch, is 933 miles; but including the Sanpu amounts to 1,750 miles. The combined drainage area of the Ganges and Bramaputra is estimated at 580,000 square miles.

4. The IRRAWADDY.—This river also rises in the Snowy Himalaya, and flows, in a southerly direction, through the heart of the Burmese territories for nearly 800 miles, till it reaches the British province of Pegu, after which it continues in a similar course for 270 miles, reaching the Bay of Bengal by several mouths, which form its delta. Although a very large river, having a course of nearly 1,100 miles, the Irrawaddy receives hardly any tributaries, and none of the smallest importance after entering British territory. Its branches in the lower part, although called rivers, have no independent sources. They are increased by torrents coming down in the wet season a few miles from the hill-sides, and thus greatly add to the volume of the water constantly pouring down from the melting snows of the mountains, where the ultimate and permanent sources exist. It was at one time supposed that the Sanpu river, the main branch of the Bramaputra, entered the Irrawaddy; but it has been proved that this is not the case.

5. The NERBUDDA.—This stream rises in the high plateau of the Vindya mountains, between 3,000 and 4,000 feet above the sea. It runs through a rugged country, with a very winding irregular course, for nearly 200 miles to Jubbulpur, where it enters the district known as the Valley of the Nerbudda. From this point

its course is nearly west, in a direct line to the sea, for 600 miles, scarcely any tributaries of the smallest importance being received, and the stream being little available for navigation.

6. The GODAVERY.—This river rises on the eastern declivity of the Western Ghats, near Nassic, only fifty miles from the shores of the Arabian Sea, and crosses the Nizam's territory in an easterly direction, receiving at intervals considerable streams, both from the north and south, and conveying the drainage of the southern slope of the Vindya range, as well as the eastern side of the Western Ghats. The principal tributaries are a group comprising the Wein Gunga, the Payne Gunga, and the Wurda, coming in from the north. The combined stream is called the *Pranhita*. The *Manjara* is a large tributary from the south. The Godavery breaks through the Eastern Ghats with an even channel, and, entering the alluvial country on the Coromandel coast, diverges into two branches, forming a delta. After running through this flat projecting tongue of land for fifty-five miles, the river enters the Bay of Bengal, after a total course of 900 miles. The delta lands are greatly enriched by periodical inundations. The total drainage area of the Godavery is about 100,000 square miles. The *Wein Gunga* rises in the Mahadeo mountains, 1,850 feet above the sea. Running to the east and south for 140 miles, it enters the Nagpur territory, and, at 340 miles from its source, it receives the Wurda. Its total course is about 440 miles. The *Wurda* is a stream having a total length of 250 miles, receiving, about 180 miles from its source, the Payne Gunga, a river of about the same magnitude, and conveying into the Wein Gunga a stream always considerable, and sometimes during the rainy season exceedingly large. It is navigable in the wet season 100 miles above its mouth. The *Payne Gunga* has a course of 320 miles, from Candeish, through the Nizam's territory, to the Wurda.

7. The KISTNA, or Krishna, rises on the eastern side

of the Western Ghats, about 4,500 feet above the sea, in lat. 18° 1'; long. 73° 41', about forty miles from the shores of the Arabian Sea. After running about 160 miles to the SE., it receives the *Wurna*, and after continuing its course for 200 miles further, it reaches the Nizam's territory, running chiefly through plains. After this it receives several tributaries, some of them of considerable importance. Its channel is generally deep, so that it cannot be used for irrigation without canals. Like the Godavery, it breaks through the Eastern Ghats to reach the alluvial plains on the Coromandel coast, which it enters by a wide estuary, not much to the south of the mouths of the Godavery. The total course of the Kistna is about 800 miles. It is hardly navigable. Among the tributaries to the Kistna are the *Beema*, rising, in the district of Poona, 3,000 feet above the sea, and having a total course of 500 miles, passing several important towns, and the *Tumbudra*, or *Tungabudra*, running 420 miles through teak forests, and capable of being employed for the navigation of rafts at all seasons.

8. The CAUVERY, a considerable river of Southern India, rises in Coorg, and forms the boundary between Coorg and Mysore, through which state it flows for some distance, running east and south-east to the neighbourhood of Trichinopoly, where it opens out into a large delta, seventy miles in length from the apex to the sea, and having a base of eighty miles, emptying into the sea by two main channels, the *Coleroon* and the *Cauvery*, of which the Coleroon is the principal. Total length of course 472 miles. The river passes from the table-land of Mysore to the low country by two falls, the upper being 370 feet, and the lower 460. During the inundations the vast body of water passing over these falls, and the fine scenery adjacent, produce magnificent views. The *Hennavutty*, the *Cubbany*, and the *Bowani* are tributaries, the latter coming from the Neilgherries.

9. The TAPTI.—A stream of some magnitude, running westward through the dominions of Sindia, in Central

India, and the territory of Gwalior, and falling into the Bay of Cambay a little south of the Nerbudda, after a course of 440 miles. It is only navigable to a short distance, and receives no considerable tributaries.

10. The MAHANUDDY.—A considerable stream, receiving the drainage of an important district on the eastern side of India, south of Bengal, and having a total course of 520 miles. It receives two tributaries of some magnitude, and enters the sea near Cuttack by a delta. From July to February it is navigable for boats for 400 miles, and during the rainy season it delivers an enormous quantity of flood water into the Bay of Bengal.

Besides these, there are many other streams of smaller importance, torrents during the rainy season, and nearly or quite dry at other times. Among them are (11) The *Braminy*, a small stream rising in the district of Palamow, and running south and east through the states of Orissa to Cuttack, entering the Bay of Bengal, after a course of about 400 miles, near Point Palmyras. (12) The *Pennar*, called generally the Northern Pennar. A considerable stream, rising in Mysore, receiving several tributaries, and conveying a large body of water into the Bay of Bengal, a little north of Madras, after running 355 miles. (13) The *Palar*, another stream rising in Mysore, and running into the Bay of Bengal, after a course of 225 miles. (14) The *Myhi*, or *Mhye*, rising in Malwa 1,850 feet above the sea, and running 350 miles into the Gulf of Cambay. It is very shallow, though never stagnant. (15) The *Luni*, in Western India, entering the Runn of Cutch, after running south-west 320 miles. (16) The *Sabur Mutti*, in Guzerat, entering the Gulf of Cambay, running south 200 miles. (17) The *Bunass*, entering the Runn of Cutch, after running south-west 180 miles.

Lakes and Lagoons.—India has very few lakes, and none of them are of any magnitude, proportioned to the lofty mountain chain forming its northern boundary.

There are, however, on and near the coast, a number of lagoons, some of them large, and not without importance in reference to inland navigation. Most of them are of water either permanently salt, or only rendered brackish during the rainy season. By far the largest lagoon is the *Runn of Cutch*, a singular expanse of low flat land, sometimes overflowed, described as part of the territory of Cutch, to which it properly belongs.* The lagoons, called "*backwaters*," on the Malabar coast, have a total length of 200 miles, and are also interesting physical phenomena belonging to the district in which they occur.† On the east coast there are a number of lagoons. *Chilka*, on the Orissa coast, measures forty-two miles in length, and is fifteen miles across in some places.‡ It is salt. *Lake Pulicat*, on the coast of the Carnatic, is thirty-three miles by eleven, and is also salt. *Lake Colair*, in the Northern Circars, is a shallow sheet of water, about 160 square miles in area. All these are near the coast. In Rajputana are three sheets of water, one of them, *Sambur*, is fifty miles in circuit, and is salt. It is dried by evaporation during summer, and the salt removed for sale. The two others (*Deedwana* and *Sirr*) are smaller. In Sind there is a small lake, and in Cashmere is *Manasa Bul*, said to be one of the most beautiful lakes in existence. There are three beautiful lakes beyond the Himalaya, and a small one at Nyni Tal.

Climate.—The characteristic of the Indian climates may be described in one word—heat. Throughout a large part of the peninsula the sun is scorching for three months of the year; the wind is hot, the land is parched, the streams, except the largest, are dried up. The hot season commences in March, and continues till June. The rains brought by the south-west monsoon then set in, and continue with intermission till October. There is then temperate weather till the end of February.

But in so large a country as India, it must not be sup-

* *See* p. 194. † *See* pp. 223, 234. ‡ *See* p. 216.

posed that the climate is everywhere the same. Very extensive tracts consist of table-land at a considerable elevation above the sea, where the extreme heats are moderated. Other large tracts are mountain sides, rendered temperate by their great height. Other tracts, again, are low, sheltered, and covered with thick vegetation, and are unfit for human residence. There are thus delicious temperate climates in the hilly regions and on the lower slopes of the high mountains, and a region of perpetual snow, above and amongst which is no life. The highest and most extreme temperature is met with in the Northern Circars and in the Carnatic. In the table-land of Central India the climate is moderate. In the plains of Delhi, 800 feet above the sea, the ground is parched in summer, and the heat very great; but the cold of winter is also pretty severe. In Upper India it is dry, and the changes of temperature are rapid and considerable. Hot winds, like blasts from a furnace, blow during three months, and in winter there is intense cold, with a hot sun at noon. Bengal is less exposed to extreme changes. The climate is moist and warm; but hot winds are unknown in summer, and frost is equally little felt in winter. The moisture, however, is oppressive for several months. The mean temperature at Calcutta is 82°, and this is the case also in Bombay. At Madras it is 84°, and in the table-land of Mysore, at the height of 7,000 feet, it is 56°.

The monsoons greatly influence the climate of India. These periodical winds are produced by the action of the sun heating the land of India during certain seasons, and modifying the trade winds, which would otherwise blow steadily from the north-east. The trade winds are thus diverted, and become south-west winds. On the western or Malabar coast these blow from April to August, the north-east wind prevailing during the rest of the year. On the eastern coast the monsoon is later and less steady; land breezes, occasionally lasting twenty-four or forty-eight hours, sometimes interfering with the monsoon. The opening and breaking up of the monsoons are everywhere accompanied by storms and hurricanes. During the prevalence of the south-west monsoon, the south-western coast is deluged with rain, the quantity falling amounting in some places to several hundred inches. Much of Southern India is, however, ill supplied with water; and famines, the result of drought, are not unusual throughout the peninsula. Generally, the eastern coast and the eastern provinces receive but little of the rain brought by the south-west monsoon; but the heaviest recorded rainfall in the world takes place at Cherra Punji, a town in Assam, in the Cossya Hills, 4,500 feet above the sea, and exceedingly healthy. The Northern Circars exhibit some peculiarities of climate. To the north of the Godavery a westerly wind, accompanied by moderate showers, begins about the middle of June; about the middle or latter part of August the rain becomes violent and irregular, and continues so till the month of October, when the wind shifts to the north-east, and stormy weather occurs. The temperature continues moderate, with little rain, till the middle of March, when the hot season commences. South of the Godavery the climate is somewhat different. During January and February a strong wind blows along the shore from the south; and as sea breezes set in every day, the temperature is moderate. In March the west wind, blowing over a loose parched soil, produces a most oppressive degree of heat, the thermometer rising to 95° in the house, and seldom falling below 90°.

Vegetable and Animal Life.—India is very rich in natural productions of all kinds. Among the principal trees are the teak, admirable for ship-building; the cocoa-nut-tree, every part of which is valuable, the fruit serving as food, the husk for cordage, the wood for water-pipes, beams, and rafters. The bamboo is largely employed in scaffolding, and for baskets and mats. The banyan, tamarind, mango, palmyra and other palms, sandal, and ebony, all abound. The sal, the sissoo, and the babul, are useful trees. In the Himalayas are pines of various kinds and great value, of which the *deodara* is now familiar in England; besides oak, and other European forest trees. The hills are covered with forests, producing drugs of various kinds, dyes, and gums. The mahowa is a valuable tree, whose flower is exported and used as food, besides being distilled for a spirit. On the coast and to some distance inland cotton is largely grown. Rice is grown on the banks of the Lower Ganges, in the Central Provinces, in the Punjab, and all round the sea-coast of the peninsula. Millet and wheat are grown in the north-western provinces of Bengal. The peasantry of the Deccan feed on poor grains called jowar, bajra, and raggi. Extensive tracts of land are appropriated to the production of articles of export; these include, besides the articles already mentioned, sugar-cane, tea, coffee, indigo, opium, tobacco, oil seeds, flax, and hemp. Pepper and cardamoms are largely cultivated on the west coast, and ginger, capsicum, cumin, sarsaparilla, anise, benzoin, camphor, coriander, and turmeric, are common field produce.

Among vegetables are yams, sweet potatoes, bunjals, gourds, cucumbers, and the common English kinds, such as potatoes, beans of various kinds, carrots, onions, garlic, spinach, and radishes. The chief fruits are plantains or bananas, mangos, almonds, dates, tamarinds, guavas, jacks, melons, grapes, pomegranates, pine-apples, peaches, strawberries, oranges, lemons, citron, lime, figs, &c. Apples are without flavour, and pears and plums do not succeed. The cultivation of tea in Assam is very important, and rapidly increasing. Coffee is largely grown in Ceylon, the Neilgherries, and elsewhere. Innumerable flowers, sweet-scented and beautiful, spangle the fields and gardens, and several species bloom even in the least likely places. The roses of Gazipur, from which the *atta* of roses and rose-water are obtained, have a world-wide celebrity.

The zoology of India is not less interesting than its botany. The forests contain a variety of wild animals, of which the elephant is the most remarkable. The Bengal elephant is the largest. The rhinoceros, wild buffalo, and bear, all inhabit the forests. Tigers, leopards, panthers, wild boars, hyænas, wolves, foxes, and jackals, squirrel, porcupine, hedgehog, and monkeys, abound both in forest and jungle, and even infest the underwood and brushwood close to civilized and cultivated lands. Lions, though widely distributed in the north, especially in Rajputana and Guzerat, are more limited. The camel is in the sandy regions of the north-west, the wild ass in great numbers traverses the great Indian desert. Deer, of many species and varieties, are found in the mountains and forests as well as the plains. The Yak, or Tartar ox; the Indian, or humped cow; the Cashmere goat, and some varieties of sheep, are important. Birds of the parrot tribe, vultures, hawks, falcons, herons, cranes, storks, flamingoes, peafowl and other fowls, pheasants, geese, swans, pigeons, ducks, and many other kinds, are infinitely common. Reptiles are numerous and remarkable. Crocodiles and alligators, large serpents, small and poisonous serpents, tortoises, and many others, are common, some on land, some in the rivers and estuaries. Sharks infest the mouths of rivers and the sea-coast, while numerous varieties of fine kinds of fishes are common in the rivers and receptacles. Insect life is marvellously rich and varied in some parts of India.

Mineral Resources.—India is rich in minerals, and yields some of peculiar value and interest. Of stony minerals most of the gems or precious stones are found, some abundantly, some almost exclusively, others of singular beauty and excellence. Diamonds of the largest size and finest water; rubies and sapphires, and emeralds, all of extraordinary size and value, were very abundant, and were accumulated by the princes before the arrival of the English. They are now much more rare. There are still, however, localities regularly searched for these productions. A large variety of gems of less value, and an infinite variety of the commoner kinds, are to be found in many of the river beds. Of the valuable stones that can hardly be ranked as gems, the variety is also very great. Jaspers, agates, cornelians, garnets, lapis lazuli, rock crystal, and many forms of quartz, are examples. Next to these may be named beautiful and valuable marbles and excellent building stone. Salt and saltpetre are both abundantly present. Coal, the most valuable of all the earthy minerals, has recently been proved to exist in so many parts of the country, that, though not yet very largely worked, it must soon become one of the great resources, and the foundation of many industries. Among other places it is found on the southern slope of the Himalaya, in Burdwan, in Palamow, in the Nerbudda districts, in Sylhet, Assam (at Cherra Punji), Berar, Rewa (Bundelcund), Cuttack, and Cutch. Of metals, gold is found in Mysore and in many streams, but the quantity is not large; tin is abundant in the Tenasserim provinces; silver, copper, and lead are known to exist; and iron is almost everywhere. But of the useful metals, except iron, the quantity is not known to be large anywhere, and the country cannot be said to be rich in them.

Population, and Races of Men.—India includes a vast variety of different peoples, of more or less mixed race, some aboriginal, but for the most part immigrated. The less accessible parts are still inhabited by hordes of barbarians, hardly removed from the lowest forms of savage life, while the coast, the plains, and the great river valleys, have been the seat of civilization of the highest oriental type from time immemorial—certainly long before the western nations had emerged from barbarism. Owing to incursions of conquering races at different times, and the gradual, but imperfect, amalgamation that has taken place, there are thus a certain number of prominent and important groups, and numerous others altogether local. The great majority are, as they have long been, Hindus, professing some one of the various forms of

Hindu idolatry. The Mahomedans, much more modern, are very numerous and influential.

In number, the total population of the peninsula is now estimated at 200,000,000. Little dependence, however, can be placed on this estimate, as every attempt at numbering the people in the different provinces is met by opposition, arising from ignorance, from superstition, and from a determination to deceive that seems altogether impossible to overcome. The difference is so great, that, in the case of Madras, a census on one occasion made the population 600,000, and, a few years afterwards, when it was known to be increased, the number was returned at 400,000.

There are marked differences in appearance among the native races in different parts of the peninsula, and these depend partly on race, and partly, no doubt, on climate. The inhabitants of Northern and North-Western India, especially the mountaineers, are tall fine men, well developed, manly, and of more than average intellect. They often exhibit energy, and, occasionally, the higher moral qualities. They make excellent soldiers, and are ingenious in certain manufactures and handicrafts, especially in the working of metals and stone. They are usually of fair complexion. The same general description applies, with some marked exceptions, to the inhabitants of the mountain districts in the other parts of India, though they are less tall. On the other hand, the people dwelling in the valley of the Ganges, and, generally, the dwellers in the lower plains and river valleys near the mouth of the stream, are lower in stature, darker in colour, less developed physically, less sincere and honest, and less favourable specimens of the human family. The Bengalis, however, stand intellectually among the first of the Indian races. The natives of Bengal and the Deccan are remarkable for their dark colour, timid, cringing, and superstitious character. On the other hand, some of the tribes in the west, especially the people of Sind and the Banias of Guzerat, are very handsome. All the races are simple in their habits, active, and capable of enduring much fatigue; the features are delicate, the limbs well formed, the eyes expressive, the hair fine and generally black, the skin soft and polished. Of the different races, the Bengalis, though weak in body and wanting in moral courage, are most capable of considerable and rapid advance in the ordinary arts of civilized life. Their intellects are quick, and there is a certain amount of cunning natural to them which makes up for their timidity. The Mahrattas are bold, active, and industrious. The natives of the Upper Provinces are brave, generous, and humane. The following brief notice of the various peoples found in India will be useful. They are arranged alphabetically.

Armenians.—Merchants wandering through India, and settled in the principal towns, chiefly refugees from former Persian persecution. Besides Armenians are Arabs, who trade from the Red Sea and Persian Gulf.

Badaks.—Hereditary thieves, inhabiting the forests north of Oude and the banks of the Chumbul. They live a nomadic life, constantly shifting their villages.

Bajicurs.—The gipseys of India, chiefly met with in the Upper Provinces.

Banias.—Merchants in Guzerat; quiet and well behaved.

Banjaras.—Travelling merchants, living in tents, and following armies. They are sometimes called *Ludanas.*

Banras.—Tribes separated from the *Newars* of Nepaul, and following many of the customs of the *Botias.*

Bats.—Wandering minstrels, in Guzerat and elsewhere.

Batties.—Thieves from the desert between Rajputana and the Indus.

Bheels.—Rude native tribes, thought to be the original inhabitants of Central India, now inhabiting the hilly parts of Guzerat and Malwa, and also the hills along the Nerbudda and the Tipti. They are expert bowmen.

Boras.—Mahomedan traders in Guzerat and the Deccan, resembling Jews in features and character.

Botias.—Tartar tribes occupying Botan, and found among the hills as far as Kumaon. They are small in stature, quiet, and industrious, but filthy in their habits.

Charuns.—A race of carriers and cattle-dealers; they are sometimes hired for protection.

Coles.—Aboriginal tribes, inhabiting the northern part of Orissa.—They are also called *Hos*, and are semi-barbarous, but hospitable, and show a love of truth, honesty, and willingness to oblige.

Coolies.—The indigenous tribes of the Western Ghats. They are hard-working, but ferocious.

Cossyas, or *Kasyas.*—A race of tall, powerful, well-formed dark people, found in the Cossya Hills, to the east of Bengal. They are peaceful, honest, and industrious.

Daudputras.—A fair and handsome Mahomedan race, who came into India from the west, and took forcible possession of the land they now hold.

Denwars.—The husbandmen and fishermen of Western Nepaul.

Garangs.—A pastoral nomadic tribe in Nepaul, sometimes becoming miners or traders. They are Buddhists.

Garrows.—A strong, hardy, and powerful people, inhabiting the hills to the east of Bengal. They are mild, honest, and faithful, and follow agricultural pursuits.

Goojurs.—An aboriginal race inhabiting the Punjab.

Gonds.—A savage people in a part of Central India formerly called Gondwana. They are bloodthirsty, cruel, and revengeful, and resemble the African negroes. They are armed with axes and arrows.

Goorkas, or *Gurkhas.*—A Mongol race dominant in Nepaul, and found also in the hill countries to the west. They make excellent soldiers, being short, sturdy fellows, hardy, brave, and enterprising.

Hos.—(See *Coles.*)

Jarejas.—The ruling class in Cutch. They are robust and warlike, but proud, cruel, and dissipated.

Jats.—A turbulent race, probably from Turkestan,

occupying a large part of the North-Western Provinces. They are short, dark, and forbidding.

Jews.—They are numerous in Western India. Those of Bombay (Beni-Israel) are of higher consideration than the others. At Cochin there are white and dark Jews, the latter regarded as slaves to the former (*see* p. 222).

Kataris, or *Katodis.*—An outcast race in the Northern Concan, held in abhorrence by the Bramins.

Katties.—Natives of Katiwar in Guzerat. A half-civilized people, formerly robbers and pirates, but now greatly reduced.

Konds.—Tribes inhabiting the central part of Orissa. They are partly civilized, and practise agriculture. They are polytheists, and till lately sacrificed human victims. They are an intelligent race, robust and muscular, and have a strong love of independence, but they are vindictive, and addicted to drunkenness.

Kotars.—An industrious race inhabiting the Neilgherries (*see* p. 215).

Kukies.—A wild tribe inhabiting the hills north-east of Chittagong. They are also called *Lunctas.*

Kurmis.—Cultivators in Cawnpur and the Doab, famed for their industry and peaceful habits.

Ludanas.—(See *Banjaras.*)

Mahrattas.—A numerous and powerful race, originally occupying the north-western part of the Deccan.

Mairs.—The inhabitants of a part of the Aravalli mountains in Rȧjputana. They are a branch of one of the original races of India, and are hereditary robbers.

Maravas.—An ancient people occupying a strip of land on the southern coast, near Cape Comorin.

Mechis.—Inhabitants of the forest portion of the Terai, from the Bramaputra to the Konkeina in Upper Assam. They are industrious and honest. In religion they are followers of Siva, but have no caste.

Moplays, also called *Mapilas.*—The Mahomedan inhabitants of Malabar. They are merchants, and are wealthy and intelligent (*see* p. 222).

Mags.—The indigenous people of Arracan. They are short in stature, and are hardy, trustworthy, and harmless.

Nairs.—The aristocracy of Malabar. They were formerly soldiers, but are now engaged in handicraft.

Namburis.—(*See* p. 236.)

Nayuks.—A wild tribe, dwelling in the jungles between the Myhi and the Nerbudda. They are idolaters.

Newars.—The original inhabitants of the fertile part of Nepaul proper, before the Goorka conquest. They are Buddhists. They are strong, peaceable, ingenious, and industrious, and chiefly engaged in agricultural work.

Oorias.—The original inhabitants of western Orissa. They are timid, effeminate, and dissolute, but honest and industrious.

Paharis.—Hill tribes, especially those inhabiting the Rajmahal Hills, and the hill country between Bagulpur and Burdwan. These are a short, thickset, sturdy people, idle and dirty, but truthful.

Parsees.—The descendants of a large colony of fire-worshippers, who left Persia and settled in Guzerat and Bombay, to avoid Moslem persecution. Many of them are now rich merchants, ship-owners, and land-owners. Others are shopkeepers, artizans, and domestic servants. They are wealthy, hospitable, and liberal, especially to their own people. They expose their dead to be devoured by birds or consumed by the elements.

Patans.—An Affgan race inhabiting Bopal, the Punjab frontier, and other parts of Hindustan.

Povindas.—A hardy race of Affgans, trading in winter and tending sheep in summer.

Rajputs.—The dominant people of North-Western India, and descendants of the ancient Hindu princes. They are tall, vigorous, and athletic, but revengeful and proud.

Ramuses.—A predatory tribe near Poona and Sattara. They believe in destiny, and have no fear of law or punishment.

Rohillas.—A people of Affgan descent, occupying the

country east of Delhi. They are tall, handsome, and intelligent, but ferocious, and not to be trusted.

Saurias, or *Sauras*.—A savage, but harmless and peaceful race, inhabiting the southern part of Orissa.

Shekawattees.—A tribe of predatory Rajputs, occupying the desert north of Jeypur.

Sikhs.—A religious community occupying the Punjab and adjacent countries. They are tall, robust, and athletic. Their occupation is chiefly warlike.

Sontals.—A primitive people inhabiting a part of the Lower Provinces of Bengal. They are very uncouth, and wear hardly any dress, but make excellent navvies. They go about armed with bows and arrows, and dwell in the wildest and least penetrable jungles.

Sudas.—A rude people, chiefly shepherds, occupying the Thur, or Great Desert, and living in a state of great privation and misery. They sell their daughters to the Mahomedans, and to the Jarejas of Cutch.

Thugs.—A confederation of murderers in Central India, greatly reduced by the severe measures of the British government. They worship the goddess Kali.

Tiars.—Cultivators on the Malabar coast, much despised by the other inhabitants.

Tudas, or *Tundavers*.—A small nomadic tribe occupying the highest valleys of the Neilgherries. They are tall, athletic, and bold, but quiet and honest. They are not Hindus, and have no recognized form of government. They have large herds of buffaloes.

Urias.—(See *Oorias*.)

Wagheas.—A predatory tribe of Rajputs in Katiwar.

Waralis.—A wild idolatrous tribe, inhabiting the jungles and forests of the northern Concan.

Yanadis.—A wild savage race inhabiting the country behind Nellore, on the Coromandel coast.

Language.—The languages used in India may be referred to four great families; (1) the Semitic; (2) the Tamulian; (3) the Indo-Chinese, and (4) the Indo-Germanic. The Semitic includes Arabic, Persian, and Turkish, none of which are strictly Indian languages, though through them an Arabic and Persian element has been introduced in certain dialects. They are therefore of comparatively small importance. The Turanian family is represented in India by a number of important languages. They are chiefly used in the south and east of the peninsula, and in the Burman empire. Thus the *Mon* dialect is spoken in Pegu; the *Gond* in the wild inaccessible country beyond Orissa, between Bengal and the Nizam's dominions; *Telugu* in Madras; *Canarese* in Mysore; *Malayalma* on the Malabar coast and Travancore; and *Tamil* on the Coromandel coast and in the south-eastern districts of the Madras Presidency. The Tamil is spoken by about ten millions of population, and its literature dates back to about the eighth century of the Christian era. The Telugu is four centuries later. It is spoken by fourteen millions. Canarese is the language of five millions, and Malayalma of two and a half millions.

The Tibetan and Indo-Chinese languages are spoken chiefly in the valleys of the Himalaya, the north-eastern provinces, and the Burman States, Assam, however, not included. Botan has a special language of this family, so also have Munipur, Tippera, and Arracan.

The Indo-Germanic is by very far the most important group of languages used in India, at least one hundred and twenty millions of persons using some one of its various forms. There are eleven distinct forms in use, besides some mixed forms that have grown up in various parts of the country. Of these, *Hindi* and *Urdu*, the latter of which is the usual official language and general medium of communication, are beyond all comparison the most important, and are spoken by more than sixty millions of human beings.

The parent of these languages is the *Sanskrit*, and its earliest known form is the language of the *Vedas*, or sacred books of the Hindus, of which the Rig Veda is the oldest. Out of this grew classical Sanskrit, which was succeeded by *Prakrit,* the form spoken by the common people, and *Pali,* the sacred language of Buddhism. The following are the chief languages of this class now spoken: (1) *Hindi,* the language of North-Western and Central Asia; (2) *Bengali,* that of the eastern provinces of Bengal. In addition to sixty millions speaking Hindi and Urdu, are twenty millions making use of Bengali. (3) *Punjabi,* spoken by sixteen millions, inhabitants of the Punjab; (4) *Mahratti,* by about ten millions of the inhabitants of Western India and the western part of the Nizam's dominions; (5) *Guzerati,* spoken by about six millions in the peninsula of Katiwar and the adjacent country; (6) *Sindi,* by two millions of the inhabitants of Sind and the Lower Indus; (7) *Cashmiri,* by three millions of the inhabitants of Cashmere; (8) *Uria,* by about two millions of the inhabitants of the Golconda and Orissa coast and the kingdom of Orissa; (9) *Pushtu,* by nearly a million of the people on the borders of Affghanistan; (10) *Parbattia,* or *Nepaulese,* by the Nepaulese; (11) *Assamese,* by the inhabitants of Assam; (12) *Beluchi,* by the inhabitants of Beluchistan.

Religions.—About four-fifths of the vast population of India—comprising upwards of a hundred and fifty millions of human beings more or less under British influence—are Hindus of some shade or other. They profess, at least, the religion of Brama, and, although subdivided into many sects, are more or less exclusively devoted to that form of worship. Braminism teaches theism, or the worship of one Supreme Being, to whom all others are subject. But worship is paid with more or less direct idolatrous feeling to *Brama,* the creative principle; *Vishnu,* the preserving principle; and *Siva,* the destroying principle. These are gods of a high order, and subordinate to them are many lesser divinities regarded as goddesses. There is a goddess of learning and eloquence (*Saraswati*); a goddess of wealth (*Lakshmi*); and the goddess *Parvati, Bowani,* or *Durga,* names representing the acting powers of evil and destruction. The Bramins have four sacred books (*Vedas*), each composed of two parts—one comprising forms of worship, and the other moral and religious instruction. They are supposed to date from about the fourteenth century before Christ. There are other sacred books (*the Puranas*), to a great extent legendary.

Out of this pure theism much polytheism has risen, the inferior principles being deified, and made the objects of direct worship. There are thus special followers of Siva and Vishnu, among whom the believers in the incarnations of Vishnu are numerous and influential. The worshippers of Siva are chiefly members of the upper classes, especially in Mysore and the Mahratta Provinces. In some large districts, the worshippers of Vishnu are the most influential, especially in his human form, as Krishna, or Kistna. The history of the incarnations (*avatars*) of Vishnu, or Kistna, is very curious, and occupies a large part of the sacred books. The Hindu religion assigns great efficacy to all religious services and the forms of devotion. The law of caste, by which all classes remain from father to son occupying the same pursuits in life, is one of the marked peculiarities of Braminism. The consideration of sacredness, as belonging to certain animals, especially the bull and cow, is another feature. The transmigration of souls, the expiation of crimes by penance, and the existence of various places of punishment and reward, are other articles of faith.

Buddhism is a form of religion much less prevalent in India now than formerly. The Buddhists do not acknowledge a Supreme Being, or rather they do not place any one God in the first rank, as greatly superior to other supposed gods. They believe in the eternity of matter, and the supremacy of intelligence as a property of matter. *Budhas* are beings (of whom there may be many) who have raised themselves by austerities of all kinds to a state of apathy, and then have evolved certain doctrines and sacred books. The last Budha lived in the sixth century before Christ. Of the two classes of Buddhists, the theists prevail in Nepaul, the atheists in Ceylon. They alike deny the authority of the sacred books of the Hindus, they do not acknowledge caste, have no respect for fire, but have great regard for animal life. They live much in monasteries.

The *Jainas*, or *Jains*, are intermediate. They agree in most points with the Buddhists, but admit caste, and worship some of the Hindu deities in addition to their own saints. Their priests are of all castes. They are numerous in Guzerat, Rajputana, and Canara. Both Jains and Buddhists use Pali as their sacred language. They date from the sixth century.

The *Sikhs*, originally pure theists, have degenerated, and regard their founder as worthy of divine honours. They have no caste, believe in transmigration, and have regard for animal life, chiefly in reference to the cow. They date from the fifteenth century, their name meaning "disciple." Their founder was one *Nanac*, the *Guru* or leader of the sect. At first they were quiet and unostentatious; but on the murder of their fourth Guru, they drew the sword, and one sect commenced to acquire a temporal power, the ruler then taking the name of *Sing*, while the rest remained quiet under the name *Kalsa*. There are other sects, and they have sacred books. They are concentrated about Amritsur and Lahore.

These are the chief modifications of Hinduism. The other principal religion of India is that of Mahomed, founded in Arabia in the sixth century, and thoroughly established a century after. There are two principal sects; the *Sunnis*, who insist on the supremacy of Mahomed over all created beings, the right succession of the four first Caliphs or successors of Mahomed, and who acknowledge tradition; and the *Shias*, who reject tradition, and regard Ali as the only rightful successor of Mahomed, and equal to him in dignity.

Of other religions, the *Parsees*, the descendants of the

ancient Ghebers or fire-worshippers, are a numerous and wealthy class. The descendants of the earliest converts to Christianity exist in considerable numbers in Malabar and the Carnatic, under the name of Syrian Christians. They date back to a very early period of Christianity, and their doctrines are very simple. There are Roman Catholics in the Portuguese and French settlements, and Protestants of all denominations throughout the country, though in very small numbers compared with the whole population.

Government.—The control of affairs throughout British India, with the exception of Ceylon, is vested in a high officer of State, entitled "the Governor-general, Viceroy of India." He is assisted by a Council, consisting of five Ordinary and some Extraordinary Members, including among the latter the Commander-in-Chief for the time being. Under the Governor-General are Governors of the Presidencies of Bombay and Madras, Lieutenant-Governors of the Lower Provinces of Bengal, of the North-Western Provinces and of the Punjab, and several Chief Commissioners. For administrative purposes, the whole country is divided into provinces, each province being governed by a Commissioner. There are two kinds of management of the provinces, one called *regulation* the other *non-regulation* (*see* p. 86). The provinces are generally subdivided into districts, over each of which is a Judge and Collector, the latter being also a Magistrate. The various native states have resident Political Agents, but are generally supreme in their own dominions. Ceylon has a Governor appointed by the Crown, subject to the Colonial department in England.

In India, the Government receives from the cultivator of the soil a certain portion of the produce as a tax. The tax is estimated and collected in three different ways. (1) *The Zemindary system.* The tax is fixed and collected from wealthy landholders, called Zemindars. It was introduced into Bengal in 1793, and prevails chiefly in the Lower Provinces of Bengal. The tax is easily collected, the revenue certain, and the landowner has the greatest interest in improving cultivation. (2) *The Ryotwary system.* The Ryots in India are the small cultivators. The tax, according to this system, which prevails in Madras and Bombay, is nominally half the produce, but really much less, and cannot exceed a certain amount. Under this system, the lands are rated according to an estimate formed of their value, and remissions are made for bad seasons. It is more difficult to collect, and less fixed in amount, but is on the whole fairer, as there is less advantage to the large landholder, who, according to the Zemindary system, is a middleman who benefits greatly on both sides. (3) *The Village system.* In this case, the assessment is made for a whole village together, and the individual payments of each proprietor is a matter concerning which the Government takes no account. It prevails in the Central Provinces, the North-Western Provinces, and the Punjab.

Revenue.—The revenue is derived, not only from the land-tax, which yields about one-half, but largely from the monopolies of salt and opium, which the Government buys of the grower and sells at a large profit. The salt monopoly, however, is abandoned in Bengal. In 1867–68, the revenue from the opium sales was nearly £9,000,000. The produce of the salt monopoly was then £5,726,093. There are also customs' dues, which yielded £2,578,632, excise £2,238,931, and stamps £2,578,632. The total revenue was £48,429,644. In 1868–9 the revenue was £49,192,007, and the expenditure (independent of guaranteed railway interest) £48,493,440.

Trade and Commerce.—The following are the principal articles of raw produce exported from British India in 1865, arranged in order of value, and the mean annual value of the exports for five years ending 1865. Since then there have been causes of disturbance, and the result is less favourable. There are now in British India neither inland, transit, nor export duties, and there is no special encouragement for shipping in British bottoms. Ceylon is not included in the following statements:—

	Year 1864–5.	Mean of five years.
Cotton	£37,573,637	£21,952,622
Opium	9,911,804	10,780,130
Rice	5,573,537	3,905,034
Indigo	1,860,141	1,855,439
Seeds	1,912,433	1,754,194
Silk	1,165,901	933,250
Jute	1,307,844	902,463
Hides and skins	725,236	796,592
Wool	1,151,022	773,221
Sugar	765,110	727,511
Coffee	801,908	555,652
Tea*	301,022	228,087

Means of Communication.—It is only within about thirty years that India has been supplied with means of communication in the interior;† and it was not till 1850 that any important progress was made in that direction. Up to that time the roads were mere tracks, the travelling of the wealthier persons being carried on in palanquins on men's shoulders, or on ponies, while the servants and tents followed on foot. Neither public coaches, waggons, nor passage-boats existed. There were no bridges, no canals, and still less were there railways or telegraphs. All these are creations of modern times, and are results of British government. There are now good roads across India, connecting all the large towns; streams have been bridged, the navigation of rivers improved, and canals cut and rendered navigable. Of late

* Nearly nine millions of pounds were exported in 1868.

† The first road of any length of which record exists is from Chittagong to Doadkundy (124 miles), open in 1810. The Grand Trunk road from Calcutta to Peshawur (1,423 miles) was commenced in 1836.

years a complete system of railway lines has been planned, and to a great extent executed. Telegraphic communication has been secured, and each year millions of pounds sterling are expended in continuing and increasing the public works, of which roads and railways form so large a part. Steamers now ply on some of the great rivers, and between every port, from Kurrachi to Rangoon. The following outline will give an idea of the present state of railway communication, and the proposed enlargement of the system.

1. East Indian Railway.—This line is completed from Howra (opposite Calcutta), by Burdwan (branch to coal-fields), Murshedabad (by a short branch), Rajmahal, Bagulpur, Dinapur, Benares, Mirzapur, Allahabad (branch to Jubbulpur), Cawnpur (branch to Lucknow), and Agra, to Delhi. The "Burdwan branch" will be a chord line connecting Burdwan with Lukheserai on the main line, crossing the coal-fields of Bengal, and saving 70 miles of distance from Calcutta to all places beyond Lukheserai. It is expected to be completed in 1870. The "Jubbulpur branch" (225 miles) will connect with the NE. branch of the Great Indian Peninsula Railway. The total length of the line and branches will be 1,501¼ miles, of which 1,356¾ are opened.

2. Oude and Rohilcund Railway.—The branch from Cawnpur to Lucknow (42 miles in length) it is proposed to connect under this name with a line reaching from Moradabad to Buxar on the Ganges, having branches to Allygur, Nyni Tal, Cawnpur, and Benares. The total length proposed is 672 miles.

3, 4, 5. Sind, Punjab, and Delhi Railways.—These are classed together, being under one management, and constituting portions of one enterprise, which is to unite the port of Kurrachi with the Punjab, and meet the East Indian Railway at Delhi. The first portion, "the Sind Railway," 109 miles in length, is completed. It connects Kurrachi with the Indus at Kotri, opposite Hydrabad. From this point there is steam navigation to Multan, where "the Punjab Railway" commences. It passes through Lahore to Amritsur (246 miles) and has been completed for some time. At Amritsur "the Delhi line" commences: its length will be 320 miles, and passes through Umballa, Saharunpur, and Meerut. 174 miles are completed from Amritsur.

6. LAHORE AND PESHAWUR RAILWAY.—From Lahore a line to Peshawur (273 miles) is in construction by Government.

7. BOMBAY, BARODA, AND CENTRAL INDIA RAILWAY.—This line runs north from Bombay, past Surat and Baroda, to Ahmedabad. It is proposed to continue it to Kaira in Guzerat, and by a line through Rajputana to Delhi. The total length at present in construction is 312½ miles, only five miles being uncompleted.

8. EASTERN BENGAL RAILWAY.—Starting from Calcutta, it passes north-east through Chinsura and Kishnagur to Kushtea, and is being extended to Goalunda. Total length, 159 miles, of which 114 are completed.

9. CALCUTTA AND SOUTH EASTERN.—A short line, 29 miles in length, connecting Calcutta with Port Canning, on the Mutla branch of the Ganges. It is worked by the Eastern Bengal Railway.

10. GREAT INDIAN PENINSULA RAILWAY.—This line commences from Bombay and runs to Calliani on the main land, a distance of 33¼ miles. Thence there are two main lines, one to the north-west and the other to the south-east. The former crossing the mountains by the Thull Ghat, passes Nassic, Bosawul (branch to Nagpur), Kundwa, and Nursingpur to Jubbulpur, where it meets the East Indian Railway. At Hurd, a branch will be constructed to Mow and Indore. The south-eastern line, crossing the Bore Ghat, passes Poona and Sholapur, and is being continued through a part of the Nizam's dominions to Raichur, where it meets the Madras Railway. The total distances are 1,266¾ miles, of which 873¾ are completed.

11. MADRAS RAILWAY.—Leaving Madras, this line first reaches Arconum Junction (42½ miles), whence there are

two main lines, one north-west (341 miles), by Tripetty, Cuddapa and Gundacul to Raichur, where it will meet the Great Indian Peninsula line. At Gundacul there is a branch to Bellary. Of this line, 185 miles out of Madras are completed, and 156 (from Gundacul to Raichur) yet unopened. The south-west line, leaving Arconum, passes by Vellore and Salem to Erode Junction. Crossing the Cauvery, it then skirts the foot of the Neilgherries to Coimbatore, after which it crosses the Western Ghats, and reaches the Malabar coast at Beypur, near Calicut. Of this line, whose total length will be 528 miles, there are 492 already completed.

12. GREAT SOUTHERN OF INDIA RAILWAY.—This line starts from Negapatam, on the Coromandel coast, and proceeds westwards to Tanjore and Trichinopoly, whence it follows the valley of the Cauvery to the Madras Railway at Erode (168 miles). It is intended to continue it through Southern India to Tinnevelly and Tuticorin.

13. CEYLON RAILWAY.—This line runs from Colombo northwards in an irregularly curved direction to Kandy, the former capital of the island in the interior.

Canals and Irrigation Works.—The physical configuration of India, and the climate that results from its geographical position, are such as to render cultivation and successful agriculture dependent almost entirely on the supply of water. Where water is to be had, and is applied in a right way, there is hardly a limit to the luxuriance of the crops of all kinds. Without an ample supply of water at the right season, the whole surface of the land is a dry cheerless waste. To prevent this there are in each of the three Presidencies many large canals whose main object is to supply water for cultivation. There are also multitudes of reservoirs (tanks they are called in India), so that some of the rivers carry but little water to the sea, the chief proportion being kept back by dams and other contrivances, and employed to fertilize the soil. In the Madras Presidency alone, there are said to be 53,000 tanks and channels in repair, and 10,000 out of repair, having probably 30,000 miles of embankments.*

The most imposing works of this character, in point of magnitude, are the canals in the Punjab and the North-Western Provinces, consisting of the Bari Doab, the Eastern and Western Jumna, and the Ganges Canals, which are 565 miles, 130 miles, 440 miles, and 653 miles in length respectively. Besides the above, there are in the Punjab the Delhi and Gurgaon Irrigation Works, and a vast number of inundation canals, chiefly in the south-western portion of the province. These latter consist of channels, taken from the rivers, which are full of water during the latter part of the spring, the summer, and autumn, and are empty during the winter. The total length of these canals, taken from the Sutlej, Chenab and Indus rivers, is 1,233 miles.

In the North-Western Provinces, besides those already named, are the Doon Canals, whose total length is 66½ miles. There are also the Rohilcund Canals, and the Agra, the Jansi, and the Humirpur Irrigation Works.

In Bombay, but few irrigation canals exist, and those only of the smallest class; but in Sind, the delta of the Indus is scored with numerous inundation canals, the principal of which are the Eastern and Western Narra, the Gar, the Bigari, the Sukkur, and Shadadpur, the Futali, the Aliwah, the Mitrow, and the Thur Canals.

In the Madras Presidency the principal canals are to be found in the deltas of the Godavery, the Kistna, and Coleroon rivers. Others take their water from the Palar, Pennar, and other rivers; there is also a very important work now approaching completion, constructed

---

* Some idea may be had of the extent of these works, when it is known that in the neglected Ponairy tank (reservoir), in Trichinopoly, the water was kept in by thirty miles of embankments, including an area of sixty or eighty square miles. The Veranum tank, still in use, has an area of thirty-five square miles, and twelve miles of embankments, and now, after enduring for a period almost fabulous, yields an annual revenue of £12,000 to the Government.

by the Madras Irrigation and Canal Company. This canal, commencing on the Tungabudra river, at Sunkasala, will give irrigation as well as navigation to the country between Sunkasala and the sea-coast at Nellore.

Among the canals at present under construction may be noticed the Sutlej Canal, which starts from the Sutlej at Rupur, re-entering the river again lower down at Ferozpur, having also branches extending into the Puttiala State, and one which will join it with the Western Jumna Canal. The Agra Canal will be an extension of the Western Jumna from its terminus at Delhi to Agra. The Sone Canal project consists of a series of canals, deriving their water from the river Sone, and falling on either side of that river into the Ganges, extending in their range as far as Patna on the one hand, and Chunar on the other. Finally, a complete network of canals will furnish the means of irrigation to the provinces of Midnapur, Orissa, and Cuttack, extending from Midnapur in the north to the Chilka Lake, near Ganjam.

Political and Natural Divisions.—India is naturally divided into several regions by its systems of mountains, table-lands, and river valleys. These do not exactly correspond with the political divisions, although they agree in a very great measure. The following arrangement and correlation may be useful:—

NATURAL DIVISIONS.	POLITICAL DIVISIONS.
Mountain system of the Himalaya Lower Valley of the Ganges and Bramaputra Valley Irrawaddy Valley Burman Coast.	Lower Provinces of Bengal North - Eastern Native States Burman States.
Valleys of Upper Ganges and Jumna Valleys of the Sutlej and Upper Indus The Thur, or Desert.	North-Western Provinces Oude Punjab Rajputana and Native States.

Table Land of Malwa Northern Deccan Orissa Coast.	Central Provinces Hydrabad Central India Orissa States.
Lower Valley of the Indus Valleys of the Nerbudda Vindya Range Western Ghats.	Sind, Guzerat and Cutch Bombay Districts.
Eastern Ghats Southern Deccan.	Madras Districts Mysore and Coorg Malabar States.
Ceylon.	Ceylon.

Progress of British Power in India.—The following tabular statement may be useful to the student for reference.

1599. Charter granted by Queen Elizabeth to English merchants to trade with India.

1601. Sailing of the first expedition to Sumatra.

1602. Factory established at Bantam, in Java.

1613. Permission granted from the Mogul to establish a factory in Surat.

1634. Permission granted to trade with Bengal through Peepley, in Balasore, on the Orissa coast.

1640. Fort St. George (afterward Madras) founded.

1650. Hoogly factory erected, under privileges granted by Shah Jehan.

1661. Company's charter renewed by Charles II.

1668. Island of Bombay (given to Charles II., in 1664, by the Portuguese) transferred to the Company.

1676. Mint at Bombay established, by permission of Charles II.

1693. Charter of East India Company renewed a second time for twenty-one years.

1698. Madras, with a territory of five miles along the shore, acquired. Fort William factory established, and a new Company incorporated.

1700. Calcutta acquired from the son of Aurungzebe.

1702. The two Companies united as "The United Company of Merchants trading to the East Indies."

1711. East India Company organized in England for the political government of British possessions in India.

1716. Important privileges granted by the Mogul Emperor and eighty-three towns purchased.

1757. Battle of Plassy, securing Bengal. The "Twenty-four Pergunnas" acquired from the Nawab.

1759. Masulipatam acquired from the Nizam.

1761. The French expelled from the Deccan. Chittagong and Burdwan acquired from the Nawab of Bengal.

1765. Bengal, Behar, and part of Orissa, granted by the Emperor of Delhi, in consideration of tribute. District round Madras obtained from the Nawab of the Carnatic.

1767. Northern Circars granted by the Nizam.

1772. Nawab pensioned, and Bengal taken possession of by the Company.

1773. Government of India reorganized by Act of Parliament.

1775. Benares acquired from the Vizier of Oude.

1776. Island of Salsette taken from the Mahrattas.

1781. Privileges of the East India Company confirmed for ten years.

1784. Board of Control created.

1786. Pulo Penang taken from the King of Quedah.

1790. Large addition of territory obtained from Tippoo Sultan in Madras Presidency.

1792. Malabar taken from Tippoo Sultan. Salem ceded.

1799. Privileges of the Company confirmed for twenty years, the powers of the Board of Control being extended. Tanjore, South Canara, Coimbatore, and other territories acquired in Madras Presidency.

1800. Bellary and other places taken from the Nizam.

1802. Carnatic taken under British administration. Allahabad, Cawnpur, Bundelcund, Futtepur, and a large part of the Doab ceded by the Government of Oude.

1803. Dutch portions of Ceylon acquired. Mahratta war. Delhi taken. Shah Alum pensioned. Agra and other places ceded. Many districts ceded by Sindia. Cuttack taken.

1805. Much of the territory of the Guicowar surrendered.

1813. Privileges of East India Company reconfirmed for twenty years. Provision made for the Church of England in India.

1815. Parts of Nepaul taken. Kingdom of Kandy, in Ceylon,

secured. French possessions in India restored. The Dera Doon taken from the Goorkas.

1816. Parts of Cutch and several districts in the NW. Provinces secured.

1817. Peishwa conquered and pensioned, and territory annexed to Bombay. Large cessions exacted from Nagpur. Holkar's territory and Sindia brought under arrangement. Parts of Guicowar's territory added. Supremacy over Central Indian States secured. Cutch subsidized.

1818. Candeish and Malwa acquired from the Holkar family, and Ajmere from Sindia. Poona, Concan, and South Mahratta annexed. Districts on the Nerbudda, Sumbulpur, and Patna annexed. Bopal taken under British protection.

1820. Lands in Southern Concan taken from the Rajah of Sawunt Warri.

1824. Chittagong brought under British authority. Singapore taken from the Raja of Johore.

1825. Malacca acquired from the Dutch.

1826. Tenasserim provinces and Arracan conquered. Assam secured.

1834. Coorg taken.

1838. Upper Assam taken under British control.

1840. Kurnul taken from the Nawab.

1843. Sind taken from the Ameers of Sind.

1845. Cashmere taken. Serampur and Tranquebar purchased of the Danish Government.

1848. Sattara lapsed.

1849. Punjab annexed.

1852. Pegu and Martaban finally occupied.

1853. Nizam's dominions regulated. Nagpur annexed. Berar assigned. Company confirmed in their possession only during pleasure. Council remodelled.

1856. Oude annexed.

1857. Great Rebellion.

1858. Government of India transferred from East India Company to the Crown.

1859. Punjab and other territory reorganized, and placed under a Lieutenant-Governor.

1861. Central Provinces organized.

1862. British Burma organized.

1866. Botan Dooars ceded.

CHAPTER II.

NORTHERN INDIA.—EASTERN DIVISION.

Boundaries and subdivisions :—(1) The *Lower Provinces of Bengal.* General Account.—CALCUTTA. *Regulation Provinces:* Bagulpur Province — Burdwan Province — Chittagong Province—Cuttack Province—Dacca Province—Nuddea Province—Patna Province—Rajshaye Province. *Non-regulation Provinces:* Assam—Cooch-Behar. (2) *British Burmah:* Arracan — Pegu — Tenasserim and Martaban — Andaman Islands—Nicobar Islands. (3) *Straits Settlements:* Malacca —Penang—Singapore—Wellesley Province. (4) *Native States of North Eastern India:* Botan — Munipore — Nepaul — Sikkim—Hill Tippera.

Boundaries and Subdivisions of Northern India.—The great valley of the Ganges and its tributaries, and the high lands that enclose it to the north and south, naturally form one of the principal divisions of the peninsula of India. The extension of this district towards the north-west into the valley of the Indus, and eastward into the valley of the Bramaputra, is both geographically and politically natural; and if to these districts we add the territory acquired by England on the eastern shores of the Bay of Bengal and the adjacent islands, we obtain an idea of what it is convenient to call Northern India.

The territory thus limited is still, however, enormously large, including nearly 480,000 square miles of country under British government, and needs subdivision. The western and north-western parts are, both politically and geographically, independent of the Lower Provinces of

Bengal. The country to the north-east and the states on the eastern shores of the Bay of Bengal are also distinct, and the independent states to the extreme west form another natural division. There are, then, two groups of states that we may call North-Eastern and North-Western India.

Adopting this method of arrangement, the following general grouping of the whole of Northern India will be found convenient:—

North-Eastern India.	Area in square miles.
1. Lower Provinces of Bengal . . .	176,813
2. British Burma	90,070
3. Straits Settlements	1,575
4. Native States of North-Eastern India .	95,477
North-Western India.	
5. North-Western Provinces . . .	83,690
6. Oude	22,456
7. The Punjab	95,768
8. Native States of North-Western India .	229,612
9. Native States adjoining North-Western India.	

Except those marked Native States, all these divisions belong to British India. They are under the general rule of the Viceroy of India, assisted by three Lieutenant-Governors and several Commissioners. There is also a Governor of the Straits Settlements, who acts directly under the Crown. With very few exceptions, the native states are either under British protection, or their government is influenced and controlled by a British officer, entitled the Resident, who acts as a political agent, and is responsible to the Viceroy.

1. Lower Provinces of Bengal.

General Account.—These provinces include an area of 176,813 square miles, with a population exceeding twenty millions. They reach from the extremity of the province of Pegu, in lat. 19° 15′ N., to the northern frontier of Assam, in lat. 28° 16′ N. From west to east they extend from the south-eastern boundary of the district of Mirzapur, in long. 83° 39′ E., to the western frontier of Burma. They are bounded on the north by Nepaul, Sikkim, and Botan; on the north-east by Tibet; on the east by Burma; on the south by the Bay of Bengal and Pegu; on the south-west by many petty independent states; and on the west by the North-Western Provinces. Except part of Orissa and Chittagong, they are entirely situated in the basins of the Ganges and Bramaputra, comprising the lower valley and delta of those two streams. The climate is very moist, in consequence of the heavy periodical rains, the great heat, and the numerous swamps of the delta.

The Lower Bengal provinces are nearly identical with the old provinces of Bengal, Behar, and Orissa.* Under

---

* *Bengal* was an independent kingdom in the fourteenth century, but was brought under the rule of the Emperor of Delhi (the Great Mogul) in the sixteenth century. On the breaking-up of that great empire in the eighteenth century, there was a Nawab of Bengal; and in 1765 the province was granted to the East India Company by the Emperor of Delhi.

Behar was one of the great divisions of the empire of Delhi, of which the present British district of Gaya was the southern part. The territory granted under this name included an extensive tract on both sides of the Ganges.

Orissa.—The ancient kingdom thus named comprised the district of Cuttack and part of Midnapur, in the Presidency of Bengal, and a wild region to the west, between Cuttack and Nagpur. It consists of an extensive range of high land continued from the Eastern Ghats. It is inhabited by four races, the Oorias, the Coles, the Konds, and the Saurias. (*See ante* p. 28–31.) The climate is very unhealthy, and the country is much infested with wild beasts, which include some of the largest, the rarest, and the most interesting in India. Orissa has lately become a district of painful interest, owing to the terrible famine of 1868. Being one of the old divisions of the country no longer recognized politically, it can only be referred to in this incidental manner.

the supreme authority of the Viceroy of India, they are subject to the direct control of a Lieutenant-Governor and Council. The revenue collected in 1867–8 amounted to £3,721,062. The value of the commerce (chiefly carried on with the United Kingdom, France, and China) amounted, in 1865–6, to—imports £23,000,000; exports, £28,000,000. The ports are, Calcutta, Chittagong, Balasore, and Cuttack. The luxuriance of vegetation in the valley of the Ganges is nowhere surpassed. Coal, iron, and limestone are obtained from the hills, but the low lands are covered with a great thickness of black soil and sand, entirely without stones. Agriculture is the chief occupation of the people, but the manufactures are many and important.

The East Indian and other Railways, as explained in p. 39, cross the most important parts of these provinces. There are also good roads, and steamboats on the great rivers. Telegraphic communication is complete.

This territory is under the control of a Lieutenant-Governor, subordinate to the Viceroy of India. It contains eight regulation, and three non-regulation provinces (*see* note, p. 86). The native states within the district are Sikkim, Munipore, Hill Tippera, and the group called the Cuttack Mehals. The regulation province of Chota Nagpur and the Cuttack Mehals are described in another chapter, as forming part of Central India (*see* pp. 155, 176).*

Education is actively carried on under the superintendence of Government officers throughout Bengal. The number of Government schools and colleges in 1867–8 was 3,411, and that of pupils 145,142, showing a large and rapid increase. Besides these were 2,196 schools not receiving state assistance, having 65,212 pupils.

For purposes of administration, the eight regulation

---

* Bengal, as now constituted, including all these provinces and states, has a total area of 240,642 square miles, and a population of 37½ millions.

provinces are subdivided into thirty-six districts. Each province is governed by a Commissioner, under whom each district is managed by a Magistrate Collector. The non-regulation provinces form three commissionerships, with nineteen districts. Each district is controlled by a Deputy Commissioner, except the Garrow Hills, which is managed by an Assistant Commissioner.

CALCUTTA, the chief and residential city, not only of Bengal, but of British India, is situated on the left bank of the Hoogly (the western branch of the Ganges), about 100 miles from the Bay of Bengal, in lat. 22° 34′ N.; long. 88° 20′ E. It comprises about eight square miles, with a population of 377,924. It extends along the river bank from north to south about four and a half miles, and its breadth is one and a half mile. It is altogether a modern town, built since the year 1700, upon the site of a number of villages assigned to the East India Company, and its name is derived from one of these villages, where was a famous temple dedicated to Kali, the goddess of Evil. It is divided into two portions, the northern or native town having narrow dirty crowded streets and poor houses. In the other portion the houses are palaces, the streets spacious, and there are numerous public buildings. The native merchants, chiefly Hindus, but including many Parsees, are very wealthy. South of the European town is Fort William, considered to surpass every other fortress in India in strength and regularity: its foundations were laid by Clive, after the battle of Plassy. It is capable of holding 15,000 men. Opposite Calcutta, on the right bank of the Hoogly, is *Howra,* the station of the East India Railway.

The climate of Calcutta is peculiar, and is subject to great extremes of moisture and heat. The different months have been thus characterized:—*January.* Air serene and cold. Winds N. and N.W.; fog in early morning, and heavy dews at night. Therm. min. 47°; max. 75°; mean, 66°. *February.* Pleasant and cool till the middle; wind then changes to S. and SE. Therm. 65° to 82°; mean, 69°. *March.* The hot season begins: the sun is powerful and the days warm. Strong winds from the south. Storms from the NW. towards middle and end, accompanied by violent gusts, with clouds of dust, followed by torrents of rain. Therm. 73° to 86°; mean, 80°. *April.* South wind moderating the heat till the 20th, when the wind becomes hot. Thunderstorms and rain. Therm. 78° to 91°; mean, 85°. *May* very disagreeable. Air close, still, and oppressive. Nights very sultry. Wind light, and from south, but storms frequent, with thunder and rain. Therm. 81° to 93°; mean, 85°. *June to September.* This is the rainy season. In the second week of June the wind veers round to the east, and after several days of close muggy weather, clouds accumulating constantly, and thunder heard in the evening, the rains commence, and continue, with little intermission, till October. The atmosphere during these months is cooler, and the weather is pleasant, but the damp is extreme, and everything gets mouldy. Therm. 77° to 88° or 90°. Mean, in June, 83°; in July, 81°; August and September, 82°. *October* is a variable month. The rains are breaking up and the winds changing. The days are sultry, but the mornings and the evenings are cool, the air becomes clear, and night dews recommence. Mean temp. 79°. *November.* Delightfully fair and pleasant. Cold sharp winds blow from the north or west. The air is dry, clear, pure, and calm, with no clouds. Therm. in shade, 66° to 86°; mean, 74°. *December.* Middle of the day clear and fine, but fogs at night, and early morning hot and disagreeable. North and west winds prevail, blowing sharply. Therm. 56° to 78°; mean, 66°.

Alipur, four miles SE. of Fort William, is rather a large village than a town, and contains the official residence of the Lieutenant-Governor of Bengal. It is healthy and dry. *Dum Dum,* eight miles NE. of Calcutta, is a town in which is a cannon foundry. It was formerly head-quarters of the artillery, and is now a large station, where 2,000 men can be accommodated. There is an arsenal and laboratory for the instruction of gunners. At *Barrackpur*

(16 miles to the N.) is the country house of the Viceroy, and a large military cantonment. *Mutla*, or *Port Canning*, is a new town, 29 miles south of Calcutta by railway, on the Mutla branch of the Ganges.

Regulation Provinces.

Bagulpur.—A province south of the eastern extremity of Nepaul, in the middle of the Ganges Valley, comprising a wide tract, reaching from the foot of the Himalaya to the Ganges, and thence to the hilly tracts to the south. Its area is about 17,000 square miles, and its population is estimated at 5,000,000.

Bagulpur (lat. 24° 17′—26° 20′; long. 86° 15′—88° 3′; area, 5,804 square miles.—A large district (the central portion of the province) partly flat, partly hilly, crossed by the Ganges and the East Indian Railway. A narrow strip of about one-fourth of the area lies to the north of the river. A large part is covered with impenetrable jungle, infested by wild beasts. Numerous torrents rush down the hills in the rainy season. Much rice is grown. Sugar-cane, cotton, and indigo are widely cultivated. Wheat and various kinds of millet yield large crops. There are manufactures in silk (Tussa), iron, copper, glass, and pottery. *Bagulpur*, the chief town, is on the right bank of the Ganges. It is a poor place, badly built, except the modern structures by the English Government. Near it are two round towers, seventy feet high, and Hindu sculptures on rocks in the bed of the river. Distance NW. from Calcutta, 268 miles. *Rajmahal* is a large and ancient but ruined town on the Ganges, on the main line of the East Indian Railway. Distance NW. from Calcutta, 196 miles.

Mongyr (lat. 24° 20′—26° 1′; long. 85° 40′—86° 50′; area, 3,593 square miles.—A small but healthy district, situated on the Ganges, in the western part of the province, and intersected by several streams. The southern part is high. It grows much rice and wheat, and by irrigation it yields abundant crops of opium, oilseeds, indigo, sugar, and tobacco. *Mongyr*, the chief town, is ancient, thriving, and prettily placed on the right bank of the Ganges. It has a large fort on a prominent rock, and is a military station. Cheap fire-arms and hardware are made here. Distance NW. from Calcutta, 304 miles.

Purnea (lat. 25° 9′—26° 37′; long. 86° 48′—88° 23′; area, 5,320 sq. m.—A level depressed tract of country, on the north side of the Ganges, adjacent Nepaul, traversed by numerous streams, affording great advantages of irrigation and water-carriage. There are many shallow ponds on the surface. It is stormy in spring, hot and dry in summer, and cold in winter. Earthquakes are common. Much rice is grown, besides maize and esculent vegetables. Tobacco and betel are also raised. *Purnea*, the chief town (pop. 50,000), is neat and well built, and is about three miles square, but much of it consists of plantations and gardens. Distance NW. from Calcutta, 283 miles.

Sontal Pergunnas (lat. 23° 40′—25° 10′; long. 86° 50′—88°; area 2,000 sq. m.).—Wild jungle, inhabited by Sontals. These pergunnas or subdistricts are chiefly in the SE. extremity of Bagulpur province, and are crossed by the East Indian Railway. Some are in the district of Beerboom.

Burdwan.—A flat expanse, liable to floods, traversed by the Hoogly river, and some of its tributaries and branches. It extends back for some distance from the river, and is crossed by many streams. The north-eastern part is crossed by the East Indian Railway and the Ranigunge branch. Area, 14,200 sq. m. Population, 6,000,000.

Bancora (lat. 22° 53′—23° 46′; long. 87°—87° 39′; area, 1,349 sq. m.).—A small district, level, but with gentle undulations, crossed by navigable streams, and containing a valuable coal-field. *Bancora*, 101 miles NW. of Calcutta, has a bazaar and spacious building for travellers. *Ranigunge*, on the Damuda, is close to the rich and valuable Burdwan coal-field. *Bishenpur* is another town in this district.

Beerboom (lat. 23° 32′—24° 40′; long. 86° 25′—88° 30′; area, 3,114 sq. m.; pop. 1,000,000).—It occupies the upper part of the province, and is hilly. It is crossed by several torrents, and yields large quantities of coal and iron ore in the hills in the southern and western parts. *Surrul* is a town near the left bank of the Aji river.

Burdwan (lat. 22° 52′—23° 40′; long. 87° 21′—88° 23′; area, 2,693 sq. m.; pop. 2,000,000).—A district crossed by numerous streams, branches of the Hoogly, and subject to inundation. It is exceedingly productive, and many of the proprietors are very wealthy. It contains both coal and iron, though much of the coal bearing this name is raised in Bancora. As much as 308 miles of embankments (bunds) have been constructed to keep out the inundations. The district is crossed by many roads and by the railway. *Burdwan* is the chief town. It is situated on the Damuda river, and is on the road from Calcutta to Benares. It is small and unimportant. Distance NW. from Calcutta, 74 miles. *Nuddea* (60 miles N. of Calcutta) is a very ancient city on the right bank of the great western branch of the Ganges.

Hoogly (lat. 22° 13′—23° 13′; long. 87° 34′—88° 30′; area, 2,007 sq. m.).—A district, low and level in the eastern part, and very fertile, but subject to ague and fever during the rains. February is cool and pleasant. There is much salt in the soil, which yields enormous crops. *Chandanagore*, a French settlement, is in this district (*see* p. 238). *Chinsura* is a flourishing town with a college. *Hoogly*, 27 miles N. of Calcutta, is showy, and has many modern buildings, a quay, and fine streets.

Midnapur (lat. 21° 41′—22° 57′; long. 86° 36′—87° 59′; area, 5,032 sq. m.).—A flat district, crossed by numerous watercourses, extremely hot in summer, but pleasant from October to February. It has a sea-coast, where salt is manufactured largely. *Midnapur*, the chief town, 68 miles W. of Calcutta, has a good and well supplied bazaar.

Chittagong.—A province, including the three districts of Tippera, Bulloa, and Chittagong. Of these, the two first lie east of the delta of the Ganges, and the latter on the northern extremity of the Arracan coast.

Bulloa and *Tippera* (lat. 22° 20′—24° 20′; long. 90° 30′—91° 12′; area, 4,629 sq. m.; pop. 1,500,000) are on the Bramaputra. The former (area 2,174 sq. m.) includes some considerable islands at the mouth of the Megna river (the combined Ganges and Bramaputra). There are towns in each, but neither is important.

Chittagong (lat. 20° 45′—23° 25′; long. 91° 32′—93°; area, 10,916 sq. m.).—This district includes fertile portions on the coast, watered by several rivers. In the interior it is mountainous, and inhabited by wild tribes. There are large forests abounding with elephants. Salt is made on the coast. *Chittagong*, the chief town, is situated on an estuary navigable for large vessels. It is foggy and unhealthy. Population, 120,000.

Cuttack.—This province lies on the north-west of the Bay of Bengal, between the mouths of the Ganges and the Madras Presidency, and has an extensive sea-board, called the Orissa coast (*see* p. 3), much of it dangerous from sand-banks derived from the Ganges. It is well watered by many streams, and includes a large lagoon (Chilka). It extends over an area of 7,600 square miles, the northern part consisting of the delta of the Mahanuddy. The interior of the country, behind the delta, is hilly. The *Cuttack Mehals* are adjacent native states, tributary to the British Government (*see* p. 176).

Balasore (lat. 20° 40′—21°; long. 86° 20′—87° 40′; area, 1,876 sq. m.).—This district is a narrow strip of land on the bay called Balasore Roads, near the Hoogly mouths, between the Subrunrika and the Dumra rivers. It is the northernmost of the three districts of Cuttack. *Balasore*, the chief town, is near the coast, on a small stream, and is the seaport of the province of Cuttack. It is provided with dry docks; but large ships cannot enter. The situation of the town is unfavourable, as it is on a low dreary plain, deformed by unsightly sand-hills. It was formerly occupied by the Portuguese. Distance SW. from Calcutta, 116 miles; N. from Madras, 730 miles.

Cuttack (lat. 20°—21° 5′; long. 85° 52′—87° 5′; area, 3,061 sq. m.).—The middle of the three districts of the province of Cuttack. It consists almost entirely of the united deltas of the Mahanuddy and Dumra. The rivers are low and swampy, and abound with alligators. The climate is very unhealthy. *Cuttack*, the chief town, occupies a commanding position, but its fortifications are ruinous. It is, however, large, and has some manufactures. The population is 40,000; distance SW. from Calcutta, 220 miles. There is a resident British chaplain.

Puri, or *Juggernath* (lat. 19° 40′—20° 26′; long. 85° 8′—86° 25′; area, 2,697 sq. m.).—The southernmost of the three districts of Cuttack. It includes the country south of the delta of the Mahanuddy. The climate is dry and healthy. The town of *Juggernath* is on the coast. It is one of the great strongholds of Hindu superstition. Every inch of it is holy, and the principal streets are composed of religious establishments. The great Temple of Juggernath rises majestically at the southern extremity to a height of 200 feet from the ground. It dates from the end of the twelfth century. For a long time the Government supported the scandalous abominations practised during the great annual procession. The town contains about 30,000 inhabitants. It is distant 250 miles SW. of Calcutta. Near *Canarac*, 19 miles NW. of Juggernath, was the celebrated "Black Pagoda," a temple of the sun. The Chilka lake, partly in this district, is one of the largest in India (*see* note, p. 216).

Dacca.—This province includes a very large tract of country on the left or eastern bank of the Bramaputra; area, 19,000 sq. m.; population, about three millions. The greater part of it is an uninterrupted level, with much jungle, but exceedingly fertile.

Backergunge (lat. 22° 2′—23° 13′; long. 89° 41′—91°; area, 4,439 sq. m.). A healthy but level tract, watered by the

Ganges, the Lower Bramaputra, and many other streams. The jungles abound with wild animals, and the soil is a rich alluvial mud, constantly shifting, and yielding rice, sugar, cotton, wheat, oilseeds, and pulse. There are many tidal creeks and lagoons. This district forms part of the Sunderbunds. *Burrisol*, the chief town, is on a branch of the Ganges, 11 miles N. of *Backergunge*, which is a town 125 miles from Calcutta. Both are small.

Cachar (area, 7,542 sq. m.).—Cachar is divided into two parts, North and South, of which the former was till lately considered a part of Assam. South Cachar is crossed by the Barak, a navigable stream. It grows coffee and sugar, and large tracts are covered by the mulberry. There are many tigers on the plains. Tea is the principal production. *Sylchar* is the principal place. *Assalu* is a town in the northern part.

Dacca (lat. 23° 12′—24° 17′; long. 90° 11′—90° 58′; area, 3,218 sq. m.; pop. 600,000).—A level depressed tract generally, the southern and most depressed part being entirely under rice cultivation. It is crossed by many streams and offsets of the Ganges. The climate is damp and unhealthy during the hot season (March to June inclusive). The mean annual temperature at noon is 79°; even during the cool season the air is often loaded with fog. Rice, sugar, betel, hemp, indigo, and cotton, are grown. *Dacca* is the chief town; it is situated on a wide stream, and is considered healthy. It is four miles in length and one and a quarter wide; but great part of it is a mass of ruins, overgrown with jungle, infested with tigers and snakes. There are, however, modern public buildings of some importance. The town was formerly celebrated for fine muslin made for the royal wardrobe of Delhi; but all its trades and manufactures are now departed. Distance NE. from Calcutta, 150 miles. There is a chaplain at Dacca.

Furridpur (lat. 23° 3′—24° 5′; long. 89° 30′—90° 15′; area, 1,634 sq. m.; pop. 850,000).—An alluvial tract, low and swampy in the southern and north-eastern parts, but rather more elevated in the north and north-western parts. It is crossed by many streams, tributaries to the Ganges, among which is the Barashi, or Chundna, always navigable. The soil is rich. Sugar and rice are the most important crops, but cotton, indigo, and oilseeds are grown. Cotton cloth is made here, and the merchants are wealthy. The town of Furridpur is large, and was formerly a resort of pirates, but now the place of residence of the Government officials. Distance, 115 miles from Calcutta.

Mymensing (lat. 24° 4′—25° 41′; long. 89° 28′—91° 13′; area, 6,710 sq. m.).—The northern part is hilly, and both there and in the south-western part along the right bank of the Bramaputra there is much jungle. The rest of the district is level, depressed, and marshy, and covered by lagoons. The climate is less unhealthy than in the southern part of Bengal, the weather being unsettled. There are three small towns.

Silhet (lat. 24° 3′—25° 12′; long. 91°—92° 38′; area, 4,981 sq. m.).—The northern part is a hilly, jungly tract, near the Cossya hills, inhabited by the wild Garrows, and on the east and south are similar tracts near the mountains of Cachar and Tipperah. The district is thus a basin, open towards the Bramaputra to the west, which is low, and subject to inundation, lasting from April to November. There are many towns and villages built on mounds. The valleys are fertile and beautiful, and crossed by several streams. The climate is damp and cool, but not healthy. There is much pasture-land. The chief place, *Silhet*, is a poor village, 260 miles NE. from Calcutta. It is the civil and military station of the province.

Nuddea, or Presidency.—This province occupies the central portion of the great delta of the Ganges, whose base rests on the sea at the head of the Bay of Bengal, and whose vertex is the point where the Jellinghi Channel leaves the Ganges. It has a total area of about 15,000 square miles.

Jessore (lat. 22° 28′—23° 46′; long. 88° 44′—89° 55′;

area, 3,712 sq. m.).—The surface is level and depressed, the climate fatal, the vegetation and wild animals exceedingly varied and interesting. The soil is generally fertile, and yields abundantly rice, indigo, oilseeds, sugar, tobacco, cocoa-nuts, areca-nuts, rye, pulse, hemp, turmeric, and many fruits. Mulberry-trees abound. The town of Jessore, seventy-seven miles NE. of Calcutta, is less unhealthy than formerly. It is the civil station, and contains important schools. There are several trading places in the district.

Nuddea (lat. 22° 49′—24° 10′; long. 88° 9′—89° 11′; area, 3,304 sq. m.).—All that has been said of Jessore applies to this district. There are ready means of water communication, but few roads. The district is populous and productive. *Kishnagur* is the seat of the civil establishment and the chief town. It is sixty-four miles N. of Calcutta. It is noted for its manufacture of fine muslins, and also for the models, in a sort of cement, of the various castes and classes of Hindus, made formerly in large quantity. There is a Government college here. *Plassy* (96 miles N. of Calcutta), a town on the left bank of the Hoogly, is celebrated for the great and decisive battle fought near here on the 23rd of June, 1757, by 650 European infantry and 150 artillerymen, with 2,100 sepoys, besides a few Portuguese, against the Soubadar of Bengal, with 50,000 infantry, 18,000 cavalry, and forty French artillerymen. The English were commanded by Clive. The Soubadar was defeated, and a foundation laid for the British occupation of India. The Ganges now flows over the field of battle. *Hurrisunkra* is another town, 102 miles N. of Calcutta.

Sunderbunds (lat. 21° 12′—22° 19′; long. 88° 5′—90° 16′; area, 6,300 sq. m.).—This is the name given to the great alluvial group of islands forming the extremity of the Ganges delta. It consists of a vast multitude of detached fragments of low marshy land, separated by narrow channels of brackish or salt water, affected by the tide, and formed by the deposition of mud brought down by the Ganges and Bramaputra. It is strictly an alluvial archipelago, and the whole area is overgrown by low wood, occasionally varied by fine timber, and inhabited more by tigers than by men. There are many wild buffaloes, wild swine, deer, and monkeys. A large part of the district is irreclaimable. Much salt is made from the sea-water on the shores, and very large tracts have been brought under cultivation, but the malaria during the south-west monsoon is very fatal; whole islands are sometimes swept away during a gale, and much of the surface is overflowed by periodical inundations. *Mutla*, or *Port Canning* (*see* p. 53), is the only town. The civil station is at *Alipur*, in the Twenty-four Pergunnas.

Twenty-four Pergunnas (lat. 21° 55′—22° 48′; long. 88° 6′—88° 43′; area, 2,536 sq. m.; pop. 700,000). So called from the territory originally ceded to the East India Company having contained twenty-four pergunnas, or subdistricts. It consists, to a great extent, of land within the delta of the Ganges, and is everywhere well watered by many rivers, all of which are branches of the Hoogly, which forms the western boundary of the tract. The northern portion is very rich, but the southern and eastern portions are covered with salt water and abound in jungles. The Sunderbunds, the swampy islets at the mouth of the Ganges, are to the east. Within the pergunnahs is the city of Calcutta. The city and the places adjacent, *Alipur*, *Dum Dum*, and *Barrackpur*, have been already described. (*See* p. 52.)

Patna.—An important province situated in the middle part of the valley of the Ganges. Its total area is 24,400 square miles, and the population is estimated at ten millions. Some of its districts are to the north, and the others to the south of the Ganges.

Gaya or *Behar* (lat. 24° 12′—25° 22′; long. 83° 25′—86° 6′; area, 5,372 sq. m.).—A well watered district, traversed by numerous rivers, and excessively sultry during early summer. The numerous torrents crossing it render the construction of roads difficult. Very fine rice is grown in large quantity. Wheat, barley, millets of various kinds, melons, gourds, and cucumbers are all abundant. The potato is successfully cultivated. The common European garden vegetables succeed well in the cold season. Opium, sugar, and cotton are important productions. Tobacco, betel, and indigo are also grown. There are several manufactures, among which are cotton, blankets, silks, carpets, tents, tape, thread, ropes, paper, torches, glass, coarse jewellery, coarse cutlery and hardware, turnery, leather, saddlery, ornaments in the precious metals, ink, soap, nitre, bricks, &c. Ardent spirits are distilled, dyeing is largely practised, perfumes are made from sandal-wood, roses, and jasmin. The town of *Behar* is a decayed place of 30,000 inhabitants. The original town is deserted, and the present is a collection of houses round its ruins. *Gaya* is the chief town.

Patna (lat. 25° 3′—25° 38′; long. 84° 45′—86° 10′, area, 2,102 sq. m.).—This district is bounded on one side by the Ganges, which is there a mile wide, and has a rapid current, and on the west and north-west by the River Sou, navigable for large craft. It is traversed by the Punpun and other streams. It is fertile and cultivated, producing excellent rice, opium, and fruits. The winters are mild, and the heat of summer great. It is well intersected by roads and is crossed by the railway. The town of *Patna* is the chief civil station, and comprises a fort enclosed by a rectangular wall, and large suburbs. It is an ancient city, and extends a mile and a half from east to west, along the right bank of the Ganges. The streets and houses are poor, but there are large markets and many mosques. The place is very hot in summer. The population is said to be nearly 300,000, chiefly Mahomedan; distance from Calcutta, by land, 377 miles, by water, 464 miles. Near it is *Bankipur*, the residence of the Government opium agent. *Dinapur*, some miles to the west, is an important military station on the right bank of the Ganges. The markets are well supplied. The barracks are very fine.

Sarun (lat. 25° 40′—27° 29′; long. 83° 55′—85° 30′; area, 6,185 sq. m.).—A level tract, with a gentle slope to the south-east, well watered, provided with some forest lands, which contain fine timber trees, and yielding large crops of sugar, grain of various kinds, tobacco, opium, indigo, and cotton. There is much salt made in some parts. *Chupra,* on the north bank of the Ganges, is the chief town, and contains many large handsome native houses, but only one street passable for carriages. The native town occupies a narrow swampy strip of low land. The civil station is outside the town. Distance from Benares, 118 miles; from Patna, thirty miles; population, 50,000. *Revelgunge* is also on the Ganges. A great fair is held here. *Aliganj* has a good bazaar. The north-east division of the district of Sarun is *Champarun,* whose chief town is *Betiya,* a populous town, ninety-five miles NW. of Patna. There is a good encampment place near, and supplies are abundant.

Shahabad (lat. 24° 30′—25° 46′; long. 83° 20′—84° 56′; area, 4,385 sq. m.).—About a third part of this district is an irregular plateau, 500 feet above the valley of the Ganges, and 700 feet above the sea. The rest is an alluvial flat, on which much cotton, indigo, opium, and betel are grown. The climate is very sultry, and the rains heavy. Where the ground is inundated periodically by the Ganges, the soil is exceedingly rich. The Ganges forms a boundary for eighty-eight miles. *Buxar* is an important town with a fort, situated in a commanding position. It is large and well built; distance from Calcutta, by land, 398 miles, by water, 566 miles. It is the principal grain mart in the collectorate, and is a railway station. *Arra* is a small town well supplied. There is a large and beautiful lake near the town. *Jehanabad* is also well supplied.

Tirhut (lat. 25° 26′—26° 42′; long. 84° 58′—87° 11′; area, 6,343 sq. m.).—An undulating and beautiful tract, with many groves, orchards, and woods on the banks of lakes and rivers. It receives its name from being bounded by three rivers, the Ganges, the Gunduck, and the Kosy. The climate is mild and moist; the thermometer range is small, and much of the district singularly healthy to Europeans. In the northern part, however, near the "Terai," at the base of the sub-Himalaya, there is much malaria and dysentery (*see* p. 100). The soil is rich, but often saturated with salt, extracted in large quantity by the natives. It is rich in fruits, among which are mango, li-chi, loquat, shaddock, guava, custard-apple, love-apple, tamarind, soap-nut, and many of those common in Europe. The country is traversed by many good roads, and there are abundant means of procuring irrigation. *Derbunga* is well supplied. *Hajipur* is at the confluence of the Gunduck with the Ganges, and is much frequented by pilgrims. *Mozuffurpur* is the capital. It has a Government school. It is on the Little Gunduck.

Rajshaye.—This large province extends over 18,000 square miles of the central part of the great valley of the Ganges, and has a population of 8,000,000. It is little more than one vast flat, subject to inundation every year, crossed by numerous streams, tributaries of the Ganges, whose beds are constantly shifting, and the surface of the land is covered by innumerable small pools, occupying hollows in the deserted channels of streams.

Bogra (lat. 24° 36′—25° 19′; long. 88° 45′—89° 48′; area, 1,704 sq. m.).—A level tract, cultivated chiefly with rice, sugar, and mulberry plantations, from which latter much silk is obtained. Hemp, cotton, and indigo are also grown. The town of *Bogra* is on the River Kuzattea. It is the seat of the civil establishment, has a bazaar, and is well supplied. Distance NE. from Calcutta, 247 miles.

Dinajpur (lat. 24° 53′—26° 38′; long. 88° 2′—89° 16′; area, 4,067 sq. m.).—This district touches the native state of Botan. Although flat, it slopes gently towards the south, and has a few eminences about 100 feet above the general level of the country; several rivers intersect it. Rice, the sugar-cane, betel, and hemp are the principal crops. The natives are miserably lodged in bamboo or mud huts. *Dinajpur*, the chief town, is clean and large, but not of much interest. It is 261 miles N. of Calcutta. *Hemtabad*, 25 miles W. of Dinajpur, was formerly a place of importance, and contains a remarkable mosque and other architectural curiosities, some of a date anterior to Moslem occupation. *Raeganj*, a small but busy place, 32 miles W. of Dinajpur, the principal mart of a large and rich district. There are many rich merchants, but the town is filthy.

Malda (lat. 24° 30'—25° 35'; long. 87° 50'—88° 30'; area, 1,655 sq. m.).—A thoroughly alluvial tract, resembling a delta, but 200 miles from the sea. The town of *Malda*, on the left bank of the Mahanuddy, is a wretched, ruined place, 191 miles N. of Calcutta. *Gaur* is a ruined city of great antiquity, on the left bank of one of the branches of the Ganges. It extended for 15 miles along the stream. A mosque and two gates are among the most striking remains.

Murshedabad (lat. 23° 48'—24° 47'; long. 87° 52'—88° 41'; area, 2,634 sq. m.).—An important district, formerly one of the centres of silk manufacture, and still manufacturing many coarse cotton fabrics, works in brass and iron, blankets, carpets, paper, mats, toys, carvings in wood and ivory, &c. It is crossed by several tributaries to the Ganges, and touched by the river itself at its widest part. The eastern part is flat, but the western hilly. The climate is unhealthy. *Bogwangola* is a pleasing and thriving place, and a great mart for grain. *Burhampore*, on the left bank of the Baghiretti, is the chief town. It was at one time frightfully unhealthy, but is said to be now much improved. It is finely situated. Distance, N. from Calcutta, 118 miles. *Jellinghi*, on the channel of that name, is of some importance. *Murshedabad* is six miles north of Burhampur, and on the same channel, the Baghiretti. It is a miserable collection of huts, but has a large population, and was once a place of great wealth and splendour, and the capital of Bengal. There is still much trade.—A new palace was built, at great cost, in 1840, for the Nawab, who is the descendant of the Nawabs of Bengal. *Cossimbazar* is an important suburb once celebrated for its silk manufactures.

Pubna (lat. 23° 34′—24° 36′; long. 88° 55′—89° 48′; area, 1,458 sq. m.).—A watery tract, traversed by the Ganges and the offsets from that river and the Bramaputra. It contains a number of shallow lakes. *Pubna* is a small town (the chief place of the district), 130 miles NE. of Calcutta.

Rajshaye (lat. 24° 6′—24° 58′; long. 88° 19′—89° 20′; area, 3,035 sq. m.).—A moist tract, crossed by numerous streams from the Himalaya mountains to the north, and subject to periodical inundations. Indigo and silk are the chief articles of export. Rice is the staple crop. *Rampur* is the chief town, 125 miles N. of Calcutta.

Rungpur (lat. 25° 16′—26° 21′; long. 88° 26′—89° 50′; area, 4,360 sq. m.).—A low district, a large part of which is inundated during the rains. The climate is peculiar. The hot winds of spring are little felt, and it is only from the beginning of June to the end of October that there is any sultriness. There is then little wind. The jungle is exceedingly rich in animals. The chief town is *Rungpur*, and is 268 miles NE. of Calcutta. It is a wretched place.

Non-Regulation Provinces.

These are Assam, Chota Nagpur, and Cooch Behar, each divided into a number of districts, placed under the authority of Commissioners, assisted by several civil officers. Assam and Cooch Behar are on the north-eastern frontier of India. Chota Nagpur is on the south-western frontier of Bengal, and belongs, geographically, to Central India (*see* p. 155).

Assam (lat. 24° 49'—28° 17'; long. 90° 40'—97° 1' area, 29,464 sq. m.; pop., about 1,000,000).—This large and important province is part of an immense plain, studded with numerous groups of hills, from 200 to 700 feet in height, and bordered on the north-east and south by lofty mountains. It is intersected in every direction by rivers, of which the Bramaputra is the principal. As many as sixty-one have been described, and there are others, smaller. It is said that, in this respect, Assam exceeds every country in the world of similar extent. The climate is superior to that of Bengal, the day heat being more moderate, and the nights being always cool and refreshing. The mean summer temperature is 80°, the winter 57°, and the annual mean, about 70°. The rains are of long continuance, lasting from March till October. Earthquakes are frequent, and, though not often serious, examples to the contrary are not unknown. The low lands are swampy, and in the rainy season flooded, though in the dry season, when cultivated, the soil yields abundant crops of potatoes. Among the objects of cultivation, tea has, for some years, been very important, and great numbers of plantations of the tea plant may now be seen. The quality of the tea grown is superior to that of China, and the quantity is rapidly increasing. Coffee also grows wild, and is cultivated. Sugar and cotton are obtained in any quantity by cultivating the red loam of the hills. Caoutchouc is obtained from the forests. Assam is very rich in mineral produce. Iron is found in the Naga Hills, and elsewhere. Coal is got in some places north of the Bramaputra. Silver occurs in the mountains. Large quantities of salt are obtained from the evaporation of brine springs. The wild animals are very numerous and abundant. The elephant is killed for its tusks. The rhinoceros, the tiger, bears, leopards, wild buffaloes, wild hogs, and game of various kinds, are plentiful, and the rivers are crowded with fish.

The population of Assam is partly Hindu, partly Mahomedan, but many wild indigenous tribes live among the mountains. The chief of these are the Abors, Borabors, and Nagas. They were long troublesome, and apparently irreclaimable, but are now quiet. They speak a peculiar dialect, or rather a distinct language, called Assamese.

The present inhabitants of Assam are the descendants of a warlike Tartar people who conquered the country and held it against the Great Mogul. At the beginning of this century the raja was expelled, and there was a period of anarchy. The whole province was taken possession of by the British, under the terms of a treaty with Ava, in 1826. Upper Assam was then placed under a native raja; but being grossly misgoverned, came in 1838 under British administration. The remainder was annexed at different times. The country has greatly improved since annexation, and is already taking rank among the finest and richest provinces of India. Assam now communicates with Calcutta by steamers on the Bramaputra, and the intercourse is very complete. Except in the hill districts, the people are civilized, and education has made great strides. There are many government schools, and the English language is generally taught. Besides the English, the American missionaries have long exerted themselves in improving the condition of the people.

Assam was formerly divided into the Upper and Lower Provinces, but has lately been rearranged. It now includes eight districts. They are *Camroop*, including *Gowhatty* (3,582 sq. m.), *Cossya and Jyntea Hills* (5,536 sq. m.), *Durrung* (2,275 sq. m.), *Luckimpur* (8,000 sq. m.), *Naga Hills* (part), (3,966 sq. m.). *Nowgong* (3,648 sq. m.), and *Seebsagur* (2,457 sq. m.).

The part of Assam including the Cossya, Jyntea, and Naga Hills is a rugged region, almost impregnable in a military point of view, and yielding coal, iron, and limestone. The country is interesting, the climate healthy and pleasant. There are several lakes, especially towards the south-east. The district is rather a sloping plateau or table-land, than a mountain chain; and there are ravines in the plateau, presenting bold precipices, and forming deep valleys. There are few peaked hills, and these are not lofty. Besides the mineral productions, cassia, lac, caoutchouc, honey, wax, and oranges are obtained. The inhabitants are not strict Hindus, as they eat beef, and have few religious notions. There are neither idols nor temples in the country. Near the villages are some remarkable stone monuments, resembling those of Stonehenge; and many peculiar rocks and stones are regarded as sacred by the natives. The hill tribes are very wild; and some of them are cannibals. They still make raids into the cultivated districts.

Cherra Punji is in the Cossya Hills district. It is 4,500 feet above the sea, and its temperature is 14° Fahr. lower than that of the plains of Bengal. The climate resembles in some respects that of England. There are frequent showers and fogs at all periods of the year, and occasionally severe thunderstorms, accompanied by hail. The rain-fall at Cherra Punji is exceptionally great. The mean fall is estimated at 300 inches; in extreme cases, 100 inches have fallen in six weeks, and nearly 600 inches in a single year. The periodical rains begin early, and continue beyond the usual period of the surrounding country. There was formerly a sanatorium at this place, now transferred to Darjeeling (*see* p. 70). *Durrung* is a large town on the right bank of the Bramaputra. *Gola Ghat* is a large mart for cotton and rice, near it are coal-beds and limestone quarries. *Gowhatty*, formerly very unhealthy, is now improved by sanitary regulations. It is prettily situated on a plain, bounded by the Bramaputra on the north, and a semicircular range of hills on the south. *Luckimpur* is a town of about 30,000 inhabitants on the frontier of Botan. *Nowgong* is a town of some importance. *Seebsagur* or *Seebpur* is a small town in the NE. of Assam. *Suduja* is above the point where the Dihong bifurcates. *Tezpur*, the ancient capital, is interesting, from the fine ruins that remain of its former magnificence.

Cooch Behar (lat. 25° 10′—26° 50′; long. 88° 25′—

91° 10′; area, including native states, 12,750 sq. m.).—A recently constructed province on the north-eastern frontier, adjoining and including portions of the native states of Botan and Cooch Behar. It comprises five districts. The whole province is governed by a Commissioner, and each district by a Deputy Commissioner, with the exception of the Garrow Hills, which has an Assistant Commissioner. The native portion of the province is under British administration.

Cooch Behar and *Garrow Hills.*—These two districts, whose total area is 5,711 sq. m., include native states, parts of them recently annexed. Cooch Behar is an alluvial level and well watered tract on the north side of the Bramaputra, which is there very wide, and studded with numerous islands. The Garrow Hills are very inaccessible, though not lofty, and form the country on the left or south bank of the Bramaputra.

Darjeeling (area 1,234 sq. m.).—A district in the extreme north-west of the province, south of Sikkim. *Darjeeling* is a hill station and sanatorium, 7,000 feet above the sea. The Darjeeling hills, to the north, are well adapted to the growth of tea and coffee. Immediately beyond these hills rise the snowy mountains of the Himalaya, and the lofty peak Kinchinjunga, once thought to be the highest point in the chain, is in view. The town of Darjeeling is situated in one of the spurs of the Sinchul mountains, on the southern side of a great hollow or basin. The mean temperature is 54°, and little exceeds that of England. The climate is delightful. The whole surrounding country is covered with forests.

East Dooars, with *Goalpara* (area, 4,378 sq. m.) and *West Dooars* (area, 1,427 sq. m.).—These districts, recently acquired by purchase from Botan, consist of a well watered country in the north of the province, with good alluvial soil, sloping gradually to the south-east. *Goalpara* lies between the hilly districts on the south of the Bramaputra and the river. It is prosperous, yielding cotton, tobacco, sugar, and mustard.

2. British Burma.

Under this name were united in 1862, for Government purposes, the kingdom of Pegu, occupied and retained after the war of 1852, the kingdom of Arracan, and the long line of sea-coast called Tenasserim. These latter were acquired by treaty, after the war of 1825–6. The whole territory lies on the east side of the Bay of Bengal, stretching southwards for a distance of 1,000 miles, with a total area of 90,000 square miles. There are also in the sea adjacent a number of islands, some near the coast, but others, as the Andaman and Nicobar groups, altogether detached. All are under the control of a Chief Commissioner, appointed under the Supreme Government of India. British Burma is divided into three provinces, and these again into twelve districts, over each of which is a Deputy Commissioner, acting as Collector and Civil Sessions Judge. The population is estimated at nearly 2,500,000, chiefly branches of the Indo-Chinese family, of whom the Burmese are the largest in number, and the most civilized. The language spoken is chiefly Burmese. The natural productions are the same in all parts: rice, tobacco, cotton, sugar, and pepper, are grown on the low grounds; timber, chiefly teak, on the hills. Sheep, oxen, and buffaloes are common. Wild animals abound.

The climate of the various countries of British Burma is much the same. The hot season is from March to June, the rainy season from May or June to October, and the cold season from October to March. The maximum temperature is about 100°, the minimum 60°, and the mean 80°. The rain-fall varies from 60 inches in some places to 190 in others, but is generally very heavy.

Arracan (lat. 18°—21° 33′; long. 92° 10′—94° 50′; area 13,484 sq. m.; pop., 445,483).—This province extends for 290 miles from north to south. The northern part is from seventy to ninety miles across, the southern not more than fifteen to twenty. The coast is skirted by many islands, the most important being Ramri, Cheduba, and Shapuri. A large part of the extended line of sea coast is rugged and rocky. Inland there are extensive flats and valleys, the latter of which are traversed by small streams, and are fertile and highly cultivated. There is also much low, marshy land, overrun with thick jungle and indented with tidal creeks, so as to render communication by land very difficult, and often impracticable. Further inland is the mountain frontier of the Yumadoang mountains, separating Arracan from Pegu, varying in height from 3,000 to 4,000 feet. One of the elevations is said to rise to 8,000 feet. Over these ranges there are several passes, one of them an excellent road. It is called the Aeng route. The inhabitants of these mountains are independent tribes. The chief rivers of Arracan are the *Myu*, the *Kuladen*, or Arracan river, the *Lemyo*, the *Talak*, and the *Aeng*. The three former flow southwards, nearly parallel, and at distances of about twenty miles apart. There are no lakes in the country. Along the coast and in the islands are mud volcanoes, and earthquakes have taken place in the province. Petroleum has been found near the mud volcanoes, but not in large quantity. Iron ore has been found. The province is divided into three districts, *Akyab*, *Ramri*, and *Sandoway*. The first is low and flat, and is a valley parallel to the sea coast; Ramri includes the islands; Sandoway is mountainous, and more healthy than the rest. The aboriginal inhabitants are called Mags, a Burmese tribe, who form about half the population. The town of *Aeng* is an important place on the road to Ava, reached from the sea by a navigable river. The surrounding country is fertile. *Akyab* is a sea-port town on an island at the mouth of the Kuladen. This is the most important town in the province, and is well situated for trade, which is carried on with great activity. The houses are well built, and the streets regular. Population 15,000. *Arracan*, the ancient capital, is a straggling town, inconveniently situated, fifty miles from the sea. The houses are raised several feet above the ground, to protect them during inundations. It contains the ruins of a fort, and has a good bazaar. The climate is unhealthy, and there is much trade with Calcutta. *Kyuk-Phyu* is an important military station on the Isle of Ramri, well built, thriving, and well situated on a noble harbour. *Sandoway* is on a tidal creek of the same name, ten miles from the sea.

Pegu (lat. 15° 49′—19° 30′; long. 94° 11′—96° 55′; area, 34,000 sq. m.; pop., 600,000).—One of the richest and most fertile provinces of the Burman empire, annexed to the British empire after the war of 1852, in consequence of indignities and injuries offered to the English by the Burmese authorities. It occupies the country within a considerable distance of the mouths of the Irrawaddy and the streams connecting with it at various parts of its course. These are very numerous, but there are four more important than the rest. The River Irrawaddy itself reaches the frontier of Pegu after having traversed 800 miles of country from the snowy range of the Himalaya, and runs for 270 miles through Pegu. The lower part is a magnificent alluvial delta, penetrated by a vast number of tidal creeks, and extending over 10,000 square miles. Both the Irrawaddy and some of its branches are navigable, and irrigate the country through which they pass. *Pegu* is the former capital. It is comparatively modern, being founded on the ruins of the old town, destroyed in 1757. It is on the Pegu river, and is well built, with wide streets. *Bassein*, on a branch of the Irrawaddy, has considerable trade, and commands the river, which is navigable for ships of the largest burden. It is the centre of a populous agricultural district, and the head-quarters of an American mission; population 25,000. *Dalhousie* is a modern town at the mouth of the river of the same name. *Meaday* is a frontier town. *Prome* is a large and important place; population 23,000. *Rangoon* is the most important town in Pegu, on the Rangoon river, a branch of the Irrawaddy. It has much trade, and increases rapidly; population 63,000. Large quantities of a valuable kind of bitumen, obtained from the neighbourhood, are shipped from thence

It has been frequently rebuilt. *Sitang* and *Tounghu* are towns rapidly increasing under British rule.

Tenasserim and Martaban (lat. 10° 48′—18° 25′; long. 96° 35′—99° 30′; area 30,000 sq. m.).—The province thus named extends along the eastern coast of the Bay of Bengal. It is a narrow strip of land, with a long line of bold and rocky sea-coast, in many places bordered by numerous islands, and indented by many creeks and small streams, besides several rivers of some magnitude, affording good anchorage. In the central part, between Ye and Tavoy, there are large tracts of swampy land called Sunderbunds, where the mangrove and other trees which thrive in brackish water grow luxuriantly, and entirely conceal the land. The interior of the country is a wilderness of thickly wooded hills, interrupted by long narrow valleys running from north to south. At intervals, and towards the north, there are large alluvial plains, highly fertile, watered by the rivers Salween and Sitang. These rivers are tidal, wide and deep at their mouths, and navigable for some distance from the sea. On the hills are forests of gigantic and valuable timber. In the jungle and in the forests are numerous wild animals. Elephants, rhinoceroses, and tigers are especially large. The mineral wealth of Tenasserim is very great. Besides coal of excellent quality, and iron in large quantity, tin has been found generally diffused in the soil throughout the country, chiefly as a black sand. This is the case especially at Mergui. Gold has been found in the streams of the Tavoy district, and copper in some of the islands. The climate on the coast has been considered very healthy.

The principal town is *Moulmein,* or Maulmain. It is on a small peninsula at the mouth of the Salween. There is a good port, and the town possesses wide streets, quays, and fine markets. It is well drained, and healthy. Vast quantities of teak from the forests are brought here, and there are shipbuilding establishments to utilize it. The population is 70,000. Other towns are *Amherst,* also an emporium of the timber trade. It is recently founded, and has a good harbour, difficult of access. *Attaran*, on a river of the same name, falling into the Moulmein. *Martaban*, on the Salween river, opposite Moulmein. *Mergui*, on the principal mouth of the Tenasserim river. It is healthy, and has much foreign trade. The harbour is accessible to ships of small burden. Off Mergui is a group of islands called the Mergui Archipelago. *Tavoy* (pop. 13,000) is on an alluvial plain, and has extensive rice cultivation in the neighbourhood. The streets are open, and the town is surrounded by a ditch, but, notwithstanding all this, it is very healthy. *Tenasserim*, formerly the capital, is now of no importance.

Andaman Islands.—A group of four large and several smaller islands of volcanic origin, about 200 miles west of the coast of the Tenasserim provinces, extending north and south parallel to the coast about the 93rd degree of east longitude, and with a total length of 140 miles. They consist of a mountain ridge, rising to 2,400 feet, the escarped side being towards the east and the slope to the west. The largest is 140 miles long, but divided into three parts by very narrow straits. Dangerous coral reefs surround the group, and a dense tropical forest covers the greater part of their surface. They are peopled by savages, who have no pursuit and no government. They worship the sun and moon. They wander from place to place, with no fixed habitations, and feed on fish and fruits. There are several excellent harbours, the best of them being *Port Blair*, where a penal colony for all India was established in 1858. These islands are singularly interesting for their zoology, several species of large land animals being apparently confined to them.*

Adjoining the Andaman Islands, between them and the Mergui Archipelago, is *Barren Island*, a remarkable active volcano. The whole island is a volcanic cone about two miles in diameter as it rises out of the sea.

---

* They have recently yielded a new large chimpanzee, and a new species of hog.

The *Cocos Islands* are two small uninhabited islands near the north-east point of Great Andaman. The largest is 6 miles long and 2 miles broad. The smaller 2½ miles long and a mile broad.

Nicobar Islands.—A group of nine large and some small islands, situated about 150 miles south of the Andaman Islands, and about the same distance from Sumatra. The largest is twenty miles long and eight across. They have been inhabited by piratical Malays, who carried on a considerable traffic in cocoa-nuts, betel-nuts, pigs, poultry, and yams. The climate is unhealthy. The Danes formed a settlement on the group in 1756, but abandoned the place in 1846. Formal possession was taken of the islands in the year 1869, by the Indian Government as a convict settlement.

3. Straits Settlements.

These settlements consist of the islands of Singapore and Penang, and a tract of land about Malacca. There is also a narrow strip on the west coast of the Malayan Peninsula. The chief authority is a Governor appointed directly under the Crown. There is a resident councillor in each settlement. Total area, 1,600 sq. m.

Malacca.—A compact territory, about forty miles in length and twenty-five in breadth, with an area of about 1,000 square miles. It yields rice, sago, pepper, jaggery, and some fruits and vegetables. The forests contain valuable timber. There are cattle and poultry, and fish is plentiful. Tin mines are worked in various places. The climate is very good. There are few hot nights, regular land and sea breezes, and a very small range of the thermometer. The monsoons are hardly felt. The town of *Malacca* is at the entrance of a small river, and built on both sides the river, connected by a bridge. It is very picturesque. Malacca was first occupied by the Portuguese, and afterwards by the Dutch. It was finally made over to the English in 1824.

Penang, or Pulo-Penang, also called *Prince of Wales Island,* is near the northern entrance of the Straits, off the west coast of the Malay Peninsula. It is fifteen miles long and twelve miles in extreme width, and its area is 160 square miles. There is a harbour formed by the Strait separating the island from the Queda coast. It has deep water, with good anchorage, and is spacious and well sheltered. It is a place of great commerce. The climate on the hills is said to be delightful. The whole island is covered with luxuriant vegetation, the shores abounding with cocoa-nut palms. The soil is light and rich, and the principal crops are cloves, on the hills, when cleared; tea, cotton, and tobacco, on the slopes; rice, coffee, and sugar-cane in the valleys. Nutmeg and the betel vine are extensively cultivated. Tin is found. *George Town* is the capital, and is the seat of Government of all the British possessions in the Straits. There are several villages, and many objects of interest.

Singapore.—An island in the Straits of Malacca, at the southern extremity of the Malay Peninsula, from which it is separated by a narrow strait, not more than half a mile wide at some points. The island is twenty-six miles in length and thirteen in its greatest breadth, and contains 275 square miles. The general surface is low and undulating, rising into rounded hills, one of which is 500 feet above the sea. The whole is covered with foliage to the water's edge. The climate, though hot, is not unhealthy. The soil is alluvial and rich, yielding sugar, cotton, coffee, nutmegs, and pepper. The town of *Singapore* is built on both sides of a river, navigable for small craft. Lat. 1° 16′; long. 103° 53′.

Wellesley Province is a narrow strip of territory on the west coast of the Malay Peninsula, opposite Penang, of which it is a dependency. It is thirty-five miles long and four wide, with an area of 140 square miles. Its surface is gently undulating, sloping towards the sea. Cocoa-nuts, rice, sugar-cane, and indigo are the chief products.

4. Native States of North-Eastern India.

Botan (lat. 26° 18′—28° 2′; long. 88° 32′—92° 30′; area, 19,000 sq. m.).—This territory lies to the east of Nepaul and Sikkim, between Assam and the mountains that form the southern slope of the Himalaya. It extends from east to west 230 miles, with a breadth of 120 miles. It is crossed by two ranges of mountain land, parallel to the great mountain chain; one (the nearest) 8,000 feet high generally, with occasional peaks as much as 16,000 feet, the other more distant and less lofty. Between the Himalaya and the first range is a high table-land, too bleak and barren to be habitable, except at the foot of the first range, where are most of the principal towns. To the east of the second range the land is level, and southwards of the lower range are the *Dooars*, or Passes, tracts of country of extraordinary fertility, whose produce once formed the chief means of subsistence of the people. These were ceded to the British in 1866, in return for an annual payment of money. Botan enjoys a variety of climate and scenery scarcely to be equalled in any other country. The cold of Siberia, the heat of Africa, and the pleasant warmth of Italy may all be experienced in a day's journey. At one view may be seen rugged barren hills and valleys covered with luxuriant vegetation, rushing mountain torrents and gentle streams, dense forests and sunny slopes, placid lakes and steep precipices, and vast masses of eternal snow stretching upwards into the sky.

The soil produces rice and millet in abundance. Sheep and ponies and a hardy breed of horned cattle are reared, and game of all kinds abound in the forests. The roads are mere tracks, through ravines which are occupied by torrents in the rainy season. The population consists of three classes: the priests, who occupy the best houses, do nothing, and live at the expense of the people; the chiefs, or "Penlows," who are the governing classes, but live in the coarsest manner; and the cultivators, who are miserably poor, and live in wretched huts. Women are treated as beasts of burden.

The country is governed, nominally, by a person called the "Durm Raja," supposed to be a divinity in human shape, but really by the "Deb Raja," who is elected by the Penlows, every three years, from their own number. The government is exceedingly bad, the population scanty, and the habits of the people filthy and immoral. The religion of the country is Buddhism. The principal town is *Punaka*, on the left bank of the Bagnee river, and ninety-six miles ENE. from Darjeeling. The houses are mostly built of wood. *Tasi-choyong* is a place of some size, on the river Gudada. *Toungsu*, on the road from Assam to Shassa, is of considerable importance. *Pusaka* is a place of great natural strength on the frontier.

Cuttack Mehals.—These native states, included in the Government of Bengal, belong geographically to Central India, and are described in p. 176.

Munipur (lat. 23° 50′—25° 40′; long. 93° 10′—94° 30′; area, 7,584 sq. m.; pop., 75,840).—A rugged mountainous country, south of Assam, between Assam and the Burman empire. It is intersected by one great valley, inhabited by a number of tribes in constant warfare with each other, roaming about from place to place, subsisting on the produce of the chase. In the central valley rice, pulse, sugar-cane, and tobacco grow luxuriantly, and the tea-plant flourishes throughout. There are many brine-springs. The soil is very fruitful, but there is little cultivation. Iron ore is found, and there are manufactures of iron and copper, the latter being chiefly worked as bell-metal for drinking and other vessels and coins. Cotton cloth is woven. The people are strong and active. All the males above sixteen perform state service. They pay no taxes, but are divided into four classes each of which serves ten days in rotation without any pay. They are brought on duty either as sepoys, cultivators, or artificers. There are no public works, except a road from the mountains of Cachar to the valley. The people are half pagans. Polygamy is common, and the women are slaves. A small tribute is paid to the British Government, and the country is governed by a Raja, and regarded as a neutral territory between British India and Burma. There is a town of the same name as the state. The British representative is termed the Political Agent, and resides in the town.

Nepaul (lat. 26° 25'—30° 17'; long. 80° 15'—88° 15'; area, 54,000 sq. m.; pop., 2,000,000).—This state extends from the valley of the Ganges on the south to the crest of the main Himalaya chain on the north, and from Kumaon on the west to Sikkim on the east. It is traversed by several considerable streams, and is divided into five parallel zones. It is a comparatively narrow strip of country, nearly 500 miles in length, and only 160 in breadth, exhibiting great diversity of surface and climate, and corresponding differences of vegetable and animal life. Along its southern border extends the "Terai," a remarkable district of unhealthy land (*see* p. 100). Beyond this is a forest region, producing a great variety of valuable timber. Beyond this, again, the country becomes hilly, and continues to rise in terraces. Still further north, these begin to assume a mountainous character, beyond and above which rises the great snowy range with Mount Everest (29,000 feet), Dawalagiri (26,862 feet), Kinchinjunga (28,156 feet), and others, the highest peaks in the world. Amongst the mountains are several habitable valleys, varying in height from 3,000 to 6,000 feet above the plains of Bengal. Of these, the valley of Nepaul proper is the largest, being twelve miles long and nine miles broad. It is bounded on all sides by lofty mountains, and its undulating surface is covered with a rich expanse of cultivated land, watered by numerous winding streams, and studded with villages and towns. It has the appearance of a lake-bed, and Hindu records describe it as having been so at some former time.

The climate of Nepaul is characterized by the widest extremes in different parts, and all degrees of temperature, from the cold of Siberia to the burning heat of the African desert, may be experienced in a day or two's journey. Generally, however, the climate resembles in some respects that of Southern Europe. The seasons are those of Upper India, but the rains commence earlier, and set in from the south-east. In the Terai putrid fever is common and very fatal, from the middle of March to the middle of November.

The mineral productions of this country are varied and important. Copper and iron mines are worked in the hills; lead is known to occur, arsenic is found, and there is good building material. The manufactures include utensils of copper, brass, and iron, and casting of bells. Iron, copper, timber, hides, rice, ginger, honey, and fruits are exported; coarse cotton, cloth, and paper are made. The population consists of Goorkas, Newars, Botias and aboriginal mountain tribes. The Goorkas, the ruling race, are Hindu-ized Tartars; the Newars are agriculturists, traders, and artisans, they have Chinese features, and are also of Tartar origin; the Botias live far up in the mountains adjoining Tibet. They and the Newars are Buddhists in religion. The language spoken by the Goorkas is a mountain dialect of Hindu, called Prabratiya. The dialect of the Newars is peculiar to themselves.

Katmandu is the residence of the Raja, and is regarded as the capital of Nepaul. It is also the headquarters of the representative of the British Government, called the Resident. It is on the east bank of the Bishnmutty river, in a circular valley, 4,800 feet above the sea. It extends for about a mile along the river bank; but its breadth is not more than a quarter of a mile.

The river is crossed by two bridges. Many of the public buildings are of wood, but there are some temples of brick, and most of the houses are of brick, with tiled roofs. The population is estimated at 50,000. The British Resident resides here. *Lalita Patun* is about six miles distant. *Goorka, Jemla,* and *Mukwanpur* are towns of some importance.

Nepaul is governed by an independent Raja. It formerly included large tracts of country that have been taken possession of by the English, but a part has been left under the old *regime.* The Goorkas appear to have conquered the country about the year 1768, and in 1790 the Nepaulese invaded Tibet with sufficient success to induce the Lama to appeal to the Emperor of China, who despatched a large force, conquered the Goorkas, and pursued them nearly to the capital. Nepaul then became a Chinese dependency. This, however, did not last long, and political relations were established before the close of the century with the East India Company. War soon ensued, as the terms of treaties were not attended to, and border invasions were common. In 1816, after some military operations that were eminently unsuccessful, the British, under Sir David Ochterlony, advanced into the heart of the country, and dictated terms which have since been observed on both sides. All the country to the west of the River Kali was added to the British dominions, and the remainder was given up to the intrigues of the native rulers. Nepaul is the largest of the independent states, and one of the very few that still remain in which British influence is not in all respects supreme.

Sikkim (lat. 27° 5′—28° 3′; long. 88° 2′—89°; area, 1,550 sq. m.; pop., 7,000).—A small mountainous tract between Botan and Nepaul. It resembles Botan in its physical features, and its productions are similar. On its northern frontier are some of the highest peaks of the Himalaya chain. *Taen-long* is the place of residence of the governing Raja. *Tasiding* is a stronghold on the top of a mountain, near Sikkim.

Tippera (Hill) (lat. 22° 48′—24° 30′; long. 91° 10′—92° 25′; area, 7,632 sq. m.; pop. not known).—A mountainous district, bounded on the north by the British districts Silhet and Cachar; on the east and south by the territory of Burma and by Chittagong; and on the west by British Tippera. It is almost covered with dense bamboo jungle, infested with elephants, deer, hogs, monkeys, snakes, and many birds. The people are Kookies, a black, undersized race, wild, and almost naked, the men being always armed with bows and arrows, spears, and other rude weapons. They live in huts constructed on scaffolds, from four to seven feet above the ground. The huts are made of bamboo slips, and thatched with grass. The people cultivate paddy, cotton, indian corn, and indigo, besides pumpkins and yams, beans and chilies. The tea-plant grows wild, and the soil is exceedingly fertile. The government is in the hands of a number of chiefs, who levy tribute on their dependents at will, and pay an annual tribute to the Maharaja.

N.B. **Cooch Behar** (area, 1,364 sq. m.), and the **Cossya** and **Garrow Hills** (area, 4,347 sq. m.), being now brought under British management, are described as forming parts of the non-regulation provinces of Bengal (*see* p. 70).

CHAPTER III.

NORTHERN INDIA.—WESTERN DIVISION.

5. *North-Western Provinces of Bengal.*—General Account—*Regulation Provinces:* Agra Province—Allahabad Province—Benares Province—Meerut Province—Rohilcund Province—*Non-regulation Provinces:* Ajmir Province—Jansi Province—Kumaon Province. 6. *Oude.* 7. *The Punjab.*—General Account—Amritsur Province—Cis Sutlej States—Delhi Province—Derajat Province—Hissar Province—Lahore Province—Multan Province—Peshawar Province—Rawul Pindi Province—Trans-Sutlej States. 8. *Native States of North-Western India:* Bahadurgur—Bawalpur—Bullubgur—Cashmere—Deojana—Furrucknugger—Gurwal—Hill States—Jujur—Loharu—Patowdi—Rajputana—Rampur—Shapura—Sikh States. 9. *Independent Native Countries adjoining North-Western India:* Afganistan—Beluchistan.

5. NORTH-WESTERN PROVINCES OF BENGAL.

General Account.—These extensive provinces reach from the Lower Provinces of Bengal on the east, to the Punjab on the west, and from Nepaul, Oude, and Tibet on the north, to Bundelcund, Malwa, and Rajputana on the south. They include about 83,690 square miles, with a population of 30,000,000. The northernmost part comprises a narrow strip of pestilential country, abounding in swamps, called the "Terai" (*see* p. 100). The great valley of the Ganges occupies the central part, and is one vast plain, gradually sloping from the west towards the east, so that at Saharunpur the level is 600 feet higher than at Allahabad. The result of this is that irrigation is gradually more and more necessary towards the west. The principal crops are cotton, rice, maize, millet, indigo, wheat, barley, oil-seeds, and tobacco, besides fruits and vegetables. Tea is grown in some places, and there is valuable timber in the hilly country. The climate is subject to great extremes, but is generally pleasant from October to March, and is not considered unhealthy. The hill countries are delightful, but the cold of winter is severe, and the heat of summer frightful. The hot westerly winds of April, May, and June are especially trying to European constitutions. In June these terminate with storm and rain. The total rainfall is not generally large.

The North-Western Provinces are under the control of a Lieutenant-Governor, appointed by the Viceroy in Council. There are five regulation and three non-regulation provinces. Each province is under a Commissioner, and the provinces are divided into districts, each of which is under a Magistrate Collector.

There is little mineral wealth in these provinces. Iron and copper are found in Kumaon, nitre is obtained from the plains, and there is sandstone in some places. The *kunkur*, an inferior kind of limestone, found in irregular lumps in the soil, is very abundant in the North-West Provinces west of the Ganges. The lumps vary in size, and sometimes form beds. All the streams which feed the Ganges on the right bank are impregnated with this lime, and those on the opposite side contain nitre, also obtained from the soil. For this reason the waters of the Ganges are often injurious to vegetation, and in some parts of its course are hardly fit either for human use or for cattle. Kunkur is very valuable for roadmaking, as it binds completely, and soon settles into a hard surface.

Many of the districts in the North-Western Provinces and in the Punjab are bounded by the great rivers of Northern India, which converge towards, and ultimately become lost either in the Ganges running eastward and south, or the Indus running west and south. The long triangular strips of land between the rivers are called in India *doabs.* They are generally alluvial and fertile, but considerably above the ordinary level of the streams, which run through ravines, except within a certain distance of the ordinary banks. These ravines are generally cut through alluvial deposits. Owing to this position, the land between the streams is dry and barren, unless artificially watered. Advantage is taken of the natural slope of the great plains of India, through which these rivers flow to the sea, to construct irrigation canals, by means of which the land, otherwise barren, becomes exceptionally fertile. Very important canals have been constructed, and are in course of construction, on several of the doabs (*see* p. 41).

The North-Western Provinces are already crossed by lines of railway, which, when the network is completed, will connect them with Calcutta on the one side and Bombay on the other, and will also open a direct communication to the Indus. Very little is wanting to complete the main lines (*see* p. 39). The other roads are good.

Telegraphic communication is secured throughout the North-Western Provinces.

*Regulation Provinces.**

Agra.—An extensive flat country between the Jumna and the Ganges (area, about 9,880 sq. m.). Though these two great rivers and their tributaries intersect most of the North Western provinces, the land greatly needs artificial irrigation, and suffers much from drought in dry seasons.

---

* In certain districts in India, power is reserved, by legislative enactment, of modifying, to any extent that may be deemed requisite, the introduction of the ordinary modes of obtaining revenue, and of conducting judicial administration. Such districts are called *non-regulation.* Those having fixed methods, on the other hand, are regulation districts or provinces.

Agra (lat. 26° 46′—27° 24′; long. 77° 29′—78° 55′; area, 1,877 sq. m.).—A level tract, intersected by several noble rivers, but suffering much from want of water, as the rivers run in deep channels. The soil is sandy; the water often brackish, and in dry years there is famine. The district contains a peculiar building sandstone, decaying readily.

The chief city is *Agra*, formerly the residence of the Mogul emperors, and till lately the seat of government of the North-Western Provinces. It is on the right bank of a branch of the Jumna, which is dry in dry seasons. The old city walls remain, and include a rectangular space of about eleven square miles; but of this, not half is now occupied. The present population is 145,000. There is one fine wide street, the houses of which are built of stone. There is an irregular-shaped fort, containing a small palace, very interesting for its decorations, and near it is the Pearl Mosque (Moti Masjid), an exquisite specimen of oriental architecture. The celebrated tomb called the Taj Mehal, is outside the city, about a mile east of the fort. The enclosure is a rectangle, measuring 964 feet from east to west, and 329 from north to south. The great central dome is 70 feet in diameter, and 260 feet high. The workmanship of the mosaics in the interior is unrivalled. Six miles north of the city is *Secundra*, a ruinous village, containing the tomb of the Emperor Akbar, one of the finest buildings in India. Distance NW. from Calcutta 783 miles; SE. from Delhi, 139. Elevation above the sea about 650 feet. *Ferozbad* (formerly *Chandwar*) is 25 miles east of Agra. It is a large and ancient but decayed town, surrounded by a wall. Population, 12,674.

Eta or *Eytuh* (lat. 27° 22′—28° 5′; long. 77° 40′—79° 15′; area, 1,400 sq. m.).—A district in the extreme north of the province, bordered by the Jumna, and crossed by the East Indian Railway. *Eta*, the chief town, is surrounded by a mud wall, and is encompassed by water during the periodical rains. It is 34 m. NE. of Mynpuri. Supplies and water are always abundant.

Etawa (lat. 26° 21′—27° 9′; long. 78° 46′—79° 49′;

area, 1,631 sq. m.).—A district in the *Doab*,* to the east of Agra. It is flat and unsheltered, and is very subject to the blasts of the hot winds of the spring months. The vegetation commences immediately after the rains, which are very heavy. The crops include wheat and barley, as well as rice, opium, sugar-cane, cotton, indigo, and millet. The town of *Etawa* is on the Jumna, the houses being on small summits, between which are deep narrow winding ravines. It is an ancient town, but has suffered decay. It is recently more prosperous. The population is estimated at 23,000. There is a cantonment a mile north-west of the town, on a desolate sandy plain. Distance N.W. from Calcutta, 710 miles; S.E. from Agra, 73.

Furruckabad (lat. 26° 46′—27° 43′; long. 78° 57′—80° 2′; area, 1,694 sq. m.).—A rich and fertile tract, a large part of which is inundated during the rainy season, and the rest sandy and sterile, but not incapable of yielding fine crops after inundation. The southern part is well wooded, and partly cultivated with sugar-cane and maize. Indigo grows wild, and cotton and tobacco are grown, though chiefly for home consumption. The town of *Furruckabad* is about three miles west of the Ganges. It is well built and healthy, and remarkably clean. It has considerable trade, and till 1824 contained a mint for the coinage supplied to the surrounding country. The population is returned at 56,300 persons. The military cantonment of *Futtegur* is three miles east of the town. It is the residence of the civil authorities, and is noted for the manufacture of tents. *Cunnouj* is a decayed town, thirty miles south-east of Futtegur, with few buildings

* It has been already explained that the tongue of flat land between two branches of a great river in Northern India is called a Doab, or *two-waters*; but the term is especially applied to the land between the Jumna and the Ganges. These two streams—the Jumna being one of the largest feeders of the Ganges—run nearly parallel for a long distance before uniting, and the land between them is pretty uniformly of the same character, and includes an area of many thousand square miles.

remaining entire, and whole mountains of unshapely ruins occupying a space of ground larger than the site of London. It was formerly one of the chief Indian cities, and is probably one of the most ancient.

Muttra (lat. 27° 14′—27° 58′; long. 77° 20′—78° 34′; area, 1,612 sq. m.).—A wide expanse of plain, with a few hills on the western frontier. It is much exposed to the burning winds of the spring and early summer. In parts the soil is good, but much of the district is arid and sandy, mixed with kunkur. The water of the wells is brackish. The population is chiefly Hindu. The town of *Muttra* is on the right bank of the Jumna. It has a large ruined fort, and was once encircled by lofty fortifications. It is regarded as sacred by the Hindus, as the birth-place of Kistna. *Bindraban*, eight miles to south-east of Muttra, is well situated, and the resort of pilgrims, owing to its association with the early life of Kistna. It has many buildings connected with Hindu worship. Near *Goverdun* are also interesting ruins.

Mynpuri (lat. 26° 54′—27° 50′; long. 78° 30′—79° 30′; area, 1,666 sq. m.).—A level, well watered tract, the Jumna forming its southern boundary, and the Kali Nuddi its western for some distance. The Ganges Canal passes it, and the East Indian Railway runs through it. The climate is very hot in summer and cold in winter. The usual crops of the Doab are produced in abundance, and of good quality. The town (of the same name) is on the Esan, and is a favourable station for troops.

Allahabad.—This division, nearly 13,600 square miles in extent, is partly in the Doab and partly south of the Jumna. It exhibits all the ordinary characters of the North-Western Provinces. There is excellent communication through it by means of fine navigable rivers, good roads and railway.

Allahabad (lat. 24° 49′—25° 44′; long. 81° 14′—82° 26′; area, 2,765 sq. m.).—About one-third of this district is in the Doab, near the confluence of the Jumna and

Ganges, and about sixty feet above the water-level at the junction. The rest is not much higher. All parts are easily irrigated, vegetation is luxuriant, the country is overspread with valuable timber, and important crops of cotton, indigo, and sugar are grown. Salt is largely manufactured and exported. The town of *Allahabad* is at the confluence of the two streams. A very strong fort 2,500 yards in circuit, and built of red stone, rises directly from the banks of the two streams. It contains an arsenal. An ancient palace, also within it, is used as officers' barracks. Allahabad is the seat of government of the North-Western Provinces, and though not noted for art or manufactures, is a place of great commerce, and is regarded as holy by the Hindus, who flock here at certain seasons to bathe in the united stream of the Ganges and the Jumna. Population, 105,000; distance N.W. from Calcutta, 498 miles; S.E. from Lucknow, 128.

*Banda** (or *Bundelcund, South division*), (lat. 24° 53′—25° 54′; long. 80° 3′—81° 35′; area, 3,030 sq. m.).—A fertile district, producing large quantities of wheat, barley, maize, millet, pulse, sugar, and indigo. The Banda cotton is well known in the market, and is largely exported. The town of *Banda* is large, straggling, and ill-built, but is a thriving place, with much trade in cotton. *Calleenjur* is a celebrated fort, situated on a rocky hill of syenite, rising out of the plains of Bundelcund. It is a table-land, between four and five miles in circuit, very strongly fortified. It now contains only a few hamlets, but was formerly important. It contains numerous temples devoted to the worship of the Hindu god Siva. There is a town at the base of the hill.

Cawnpur (lat. 25° 55′—27°; long. 79° 34′—80° 37′; area, 2,353 sq. m.).—A district terribly celebrated, in consequence of its share in the mutiny of 1857. It is situated

---

* Banda is a part of the extensive tract called Bundelcund, or the Bundela country. It is described in the account of Central India. (*See* p. 166).

within the Doab, and is a nearly level tract, watered by several streams, and crossed by the Ganges Canal. Millet, the sugar-cane, and maize thrive luxuriantly. Indigo, opium, and cotton are also grown, and the tobacco is celebrated. The chief town (of the same name) contains 110,000 inhabitants. It is a military station, and the seat of a large trade. Being situated on the Ganges, which is there half a mile wide, and navigable from the sea, it is admirably adapted for trade. During the mutiny of 1857, the principal buildings were destroyed, and a number of Englishwomen and children murdered.

Futtepur (lat. 25° 25′—26° 13′; long. 80° 12′—81° 23′; area, 1,580 sq. m.).—A district within the Doab, bounded on one side by the Ganges, and on another by the Jumna. The soil is fertile, and the climate dry, but subject to extreme variations in temperature. During spring it is represented as a "boundless garden." Beautiful groves of mangoes, tamarinds, and bananas overshadow the village pagodas, mosques, and tanks, and give an ever-varying beauty to the landscape, which is animated by pilgrims, peasantry, travellers on foot and horseback, heavily laden carts and camels. The principal place bears the same name. It is a large, ancient, and thriving town, with good houses and an elegant mosque. There are several other towns.

Humirpur (lat. 25° 7′—26° 26′; long. 79° 20′—80° 25′; area, 2,289 sq. m.).—A level tract, fertile in parts, and yielding the usual crops of the surrounding country. The cotton is very good. It is crossed by the Jumna and other streams. The climate is hot and unfavourable to Europeans, though not to natives. The principal place (of the same name) is situated on a tongue of land at the confluence of the Betwa and Jumna. It is large, consisting originally of several villages grouped together. Distance NW. from Calcutta, 575 miles.

Jounpur (lat. 25° 22′—26° 12′; long. 82° 12′—83° 10′; area, 1,555 sq. m.).—A level tract, with a gentle declivity of six inches to the mile towards the SE. There is a large Hindu population, chiefly agricultural. The town of *Jounpur* is built on both banks of the Gumti, a navigable stream, crossed by a bridge nearly three centuries old, considered to be one of the finest in India among the constructions of that date. The town contains a vast massive stone fort, said to have been built in 1370. There is a civil establishment and military cantonment. Distance NW. from Benares, 35 miles.

Benares.—A nearly flat fertile tract of country, crossed by many streams, covering an area of 18,330 square miles, stretching from the Nepaul Terai on the north, to the Bundelcund plateau on the south.

Azimgur (lat. 25° 36′—26° 24′; long. 82° 45′—84° 12′; area, 2,553 sq. m.).—A thickly peopled district, with a number of small towns, crossed by several streams and roads. The chief place (also *Azimgur*) is on a small stream, and has extensive manufactures.

Benares (lat. 25° 7′—25° 32′; long. 82° 45′—83° 38′; area, 995 sq. m.).—This small district is exceedingly level, no part of it being more than 300 feet above the sea. It is crossed by the Ganges, and its climate is comparatively cool, though the mean temperature is 77°. It is crossed in various directions by roads, the railroad, and canals, and yields all the ordinary crops of the valley of the Ganges. The town of *Benares* is situated on the left bank of the Ganges, at a point where the river is nearly half a mile wide and fifty feet deep in the dry season. The stream forms a bay indenting the front of the town, and displaying its peculiar and picturesque features to great advantage. It has been compared to Naples in this respect. The city rises from the river, spires, temples, ghats, or flights of broad steps, balconies, and lofty houses, intermixed with trees, walls, minarets, &c., all in close contiguity, giving it an air of a populous and densely built town. The streets are narrow, crooked, and crowded, and most of the business is transacted on the ghats. There may be seen at all hours thousands of people bathing, praying, preaching, gossiping, bargaining, lounging, and sleeping. Pilgrims from all parts of India visit Benares, and crowds of fakirs may be seen in all the hideousness of filth, cow-dung, disease, and deformity. The population is upwards of 200,000, a very large proportion being Hindus. The Hindu temples are wealthy and in good repair, while the mosques are deserted and in ruins. It has been described as the Jerusalem of Hindustan. It swarms with teachers of the Hindu worship, with Braminee bulls, and with devout, rich, corpulent, and active mendicants. The East Indian Railway passes close to the city. Distance NW. from Calcutta, 421 miles. A few miles above the city is *Chunar*, a fortress and a station. Good freestone is obtained hence. *Secrole*, three miles distant, is the station ot the civil officers, and a military cantonment.

Busti (lat. 26° 25′—27° 30′; long. 82° 20′—83° 25′; area, 2,804 sq. m.).—A district in the north-western part of the province between the Gagra river and the Terai, and crossed by the Rapti. The town of *Busti* has a bazaar, and is well supplied with good water.

Gazipur (lat. 25° 17′—26°; long. 83° 8′—84° 40′; area, 2,195 sq. m.).—A nearly level district, crossed by several streams, and with one large and several smaller sheets of water. No part of the district is more than 350 feet above the sea. The climate is healthy, except in autumn, when fevers are common. It is very densely peopled, almost entirely by Hindus, and chiefly by agricultural labourers. The chief place is *Gazipur*. It is a large town on the Ganges, pleasing in appearance, but in ruins. At the eastern extremity is the palace of Cossim Ali, a Nawab of Bengal, infamous for his massacre of British prisoners in cold blood. It is a very interesting specimen of Indian Saracenic architecture. Gazipur is celebrated for its rose-water. The roses are cultivated in fields. The leaves are distilled with double their weight of water, and the rose-water thus obtained is poured into large shallow vessels, and exposed, uncovered, to the open air at night. The jars are skimmed occasionally for the *atta*. It takes a thousand roses to procure a grain of the essential oil, which is worth, on the spot, about a shilling. A quart of the best rose-water is worth about the same sum. Both rose-water and the *atta* are exposed to the full heat of the sun for at least a fortnight after being bottled. The *atta* is frequently adulterated with sandal oil. Gazipur is 431 miles NW. of Calcutta, and 46 miles NE. of Benares.

Goruckpur (lat. 26° 7′—27° 30′; long. 82° 12′—84° 30′; area, 4,584 sq. m.).—A large district, level, except in the east and south-east, where are some low ridges. There are many small lakes, called *jils*, some of them six or seven miles long and three miles broad. The climate in the south is hot, but not unhealthy. In the north it is less healthy. The population is very large, exceeding three millions, and chiefly Hindu. The chief town is *Goruckpur*, situated on the left bank of the Rapti, a fine navigable river. It is large, but thinly peopled and unhealthy. Distance NW. from Calcutta, 420 miles.

Mirzapur (lat. 23° 50′—25° 30′; long. 82° 11′—83° 39′; area, 5,200 sq. m.).—This district is level and alluvial in the northern part, where it is a portion of the lower valley of the Ganges. In the southern part is a range of rocky uneven hills, connected with the Bindachal range. These are jungly and pestilential, but the general climate is represented as not more unhealthy than others adjacent. It is, however, cold in winter and very hot in summer. The population is almost exclusively Hindu, and exceeds 1,000,000. *Mirzapur* is the chief town. It is built on a ridge of kunkur, on the right bank of the Ganges. It looks imposing from a distance, but the houses are of mud. It is a great cotton mart, and is celebrated for its carpets. Population 80,000. It is a very wealthy place. Distance NW. from Calcutta, 448 miles; SW. from Benares, 27 miles. *Chunar* is a town on the Ganges, 21 miles E. of Mirzapur, remarkable for a strong fort, built on a rocky eminence from 100 to 150 feet above the stream. The space enclosed is 750 yards in length, and 300 yards wide at the widest part. The ramparts are from ten to twenty feet high, with many towers. Within it is an ancient Hindu palace. Outside the town is an interesting tomb.

Meerut.—This whole province is remarkably uniform. It covers an area of 11,000 square miles, watered by the Ganges and Jumna, and artificially irrigated by the Ganges canals. The soil is a loam, and yields good crops of wheat, barley, oats, pulses, tobacco, and European vegetables in the cold season, and rice, cotton, maize, indigo, and millet in the hot. There are some remarkably fertile strips of marshy land along the banks of the rivers.

Alligur (lat. 27°—27° 28′; long. 77° 32′—78° 47′; area, 1,858 sq. m.).—A level district in the Doab, with a prolonged elevation of the surface between the two rivers (the Ganges and the Jumna), thickly peopled, yielding wheat, barley, millet, and pulse for home use, and indigo, cotton, tobacco, and sugar for export. There is a town and fort in the district of the same name. *Coel*, the civil station, connects with the town by a fine avenue of trees, two miles in length. It was a place of some importance in the twelfth century.

Bulundshur (lat. 28° 3′—28° 43′; long. 77° 28′—78° 32′; area, 1,910 sq. m.).—A district in the Doab, resembling Alligur in its physical character and resources. The chief town is also *Bulundshur*. It is on the right bank of the Kali Nuddi, 40 miles SE. of Delhi. It is small, the population not exceeding 15,000.

Dera Doon (lat. 30°—30° 32′; long. 77° 43′—78° 24′; area, 934 sq. m.).—A wide fertile valley, at the south-western foot of the lowest and outermost ridge of the Himalaya. The Sewalik range separates it from Saharunpur. The length of the valley is 45 miles, and the breadth from 15 to 20 miles. It is drained by tributaries of the Jumna and Ganges. The mountains on the north-eastern frontier rise from 7,000 to 8,000 feet, the Sewalik range being from 3,000 to 3,500. The level of the water in the rivers is from 1,200 feet in the lower part of the district, to 1,470 feet in the higher, their difference representing the natural slope of the ground. The Sewalik range is especially interesting to the geologist, from the large quantity of fossil bones of extinct animals of the tertiary period found there. The climate of the district is peculiar, the thermometer ranging at different seasons from 37° to 101° in the day. June is the hottest month. July, August, and September are unhealthy, being the times of periodical rain. There is much jungle, and the uncultivated part of the district is covered with dense and almost impenetrable forest. Both plants and animals are interesting to the naturalist. It is said that "every English plant thrives luxuriantly in the Doon, where in March, April, and May a splendid show of English flowers is to be seen in all the gardens." Rice, maize, pulse, cotton, sugar, opium, hemp, indigo, and plantain are the native crops. It is considered that there are 100,000 acres of land adapted for the cultivation of tea, which is grown in large quantities. The principal town, *Dera*, is on the crest of a ridge 2,400 feet above the sea, in an extensive grove of mango trees. It is not very large, but contains good buildings and a handsome temple. Distance NW. from Calcutta, 974 miles. At *Gurudwara*, a large village, is an important annual fair. *Mussouri* is a sanatory station on the northern frontier, 7,000 feet above the sea; and *Landour*, another almost adjoining, a thousand feet higher. The latter is a depôt for sick soldiers. There is no level ground at either station, and most of the houses are built on terraces cut out of the solid rock. The views from them are beautiful, reaching on one side to the snowy peaks of the Himalaya, and looking down on the other to the valley of Dera, while, beyond the Sewalik Hills, the plains of Upper India are seen. There is good slate at Mussouri. All kinds of food are abundant, both at Mussouri and Landour. Distance from Calcutta, 1,060 miles.

Meerut (lat. 28° 33′—29° 17′; long. 77° 12′—78° 15′; area, 2,368 sq. m.).—A district forming part of the Doab. It is level, but with a ridge between the streams of the Ganges and Jumna, along which proceeds the line of the Ganges Canal. The climate is healthy, and favourable for vegetation. Much wheat is grown in the cold season, on the same soil which in summer produces large crops of sugar, indigo, and cotton. The population is large, but not thickly agricultural. The principal town is *Meerut*. It is about midway between the Ganges and the Jumna, but the Kali Nuddi, here a small stream, runs three miles east of the town. The walls are extensive, and there are some interesting remains of ancient architecture, but the town is dirty and badly built. The cantonment is ten miles north of the town. *Gurmuktesar* is the port of Meerut.

Mozuffurnuggur (lat. 29° 10′—29° 50′; long. 77° 6′—78° 10′; area, 1,650 sq. m.).—A level district, with a few sand-hills and a low ridge. Alluvial marshy tracts extend along the banks of the Jumna and Ganges, which run through it. These marshes (called "Kadirs") are very fertile, but unhealthy. The population is not dense, and not largely agricultural. The products are those of the rest of the Doab. The town (of the same name) is large. It is on the West Kali Nuddi, distant NW. from Calcutta 984 miles.

Saharunpur (lat. 29° 28′—30° 26′; long. 77° 13′—78° 15′; area, 2,227 sq. m.).—A level district rising towards the base of the Sewalik hills, which rise precipitously from the plain. There are two long ranges of sand-hills running from north to south, parallel with the course of the Ganges. Between one of these and the river is the *Kadir*, or marsh land, amounting to one-sixth of the whole area of the district. The climate is always comparatively cool, and it is absolutely cold from November to March. In April, however, there are hot winds. The soil is loam, with kunkur. The crops are those common throughout the Doab. Irrigation is supplied by the Doab Canal. *Saharunpur* is the principal town. It is on a small stream, in an open level country, surrounded by groves of mangoes and palms and enclosures of cactus. It possesses a botanic garden, which was under the care of the eminent botanists, Dr. Royle and Dr. Falconer, in succession. Great success has been obtained in the naturalization in the open air of the productions of various countries. The garden is of considerable size, containing nearly fifty acres of ground, and is tastefully laid out with walks and carriage-drives. Population, 40,000. Distance NW. from Calcutta, 1,007 miles. Height above sea, 980 feet. *Hurdwar*, or *Gangadwara*, also *Goupela*, is on the west bank of the Ganges, and a celebrated place of pilgrimage. The bathing commences on the 10th of April, and a great fair is then held. Every twelfth year the fair and resort are very much greater than usual; as much as two millions of people assembling on these occasions. Hurdwar is an important stud depôt. Distance from Calcutta, 924 miles. *Rurki* has a large Civil Engineering establishment.

Rohilcund.—An extensive tract lying to the east of the Ganges, between the Doab and the mountains. It includes an area of about 11,700 square miles. In its physical characteristics it resembles the Doab. The soil is fertile, and where there is jungle it is dense and impenetrable. The Rohilla Patans or Rohillas occupied this province during the last century.

Bareilly (lat. 28° 2′—29° 19′; long. 79° 4′—80° 12′; area, 2,925 sq. m.).—A level district with a fine climate, colder in winter than might be expected from the latitude, and suffering little from hot winds. The soil is fertile and highly cultivated. The population is nearly a million and a half. *Bareilly* is the chief town. It is pleasantly situated, 788 miles from Calcutta, and is a considerable though decayed place, with much trade and some manufactures. *Futegung* is a thriving populous town, 12 miles NW. of Bareilly.

Bijnour (lat. 28° 54′—29° 58′; long. 78° 1′—78° 53′; area, 1,884 sq. m.).—A district chiefly remarkable for the successful culture of the sugar-cane. It is watered by two streams, the Koh and the Ramgunga. The chief town is *Bijnour.* It is not large, but is the seat of government. Distance NW. from Calcutta, 800 miles.

Budaon (lat. 27° 38′—28° 39′; long. 78° 21′—72° 35′; area, 1,971 sq. m.).—A low, level, and generally fertile district, watered by the Ganges and other streams. It is thickly peopled, almost entirely by Hindus. Its chief town is *Budaon.* It has a population exceeding 20,000.

Moradabad (lat. 28° 15′—29° 27′; long. 78° 10′—79° 24′; area, 2,461 sq. m.).—The northern and north-eastern parts of this district are 1,000 feet above the sea, and the country slopes gently from that direction to the south at the rate of ten feet in a mile. Near the base of the hills is marsh land (Terai), with many springs and streams, luxuriant vegetation, and deadly malaria. There are marshes, also, near the Ganges, but these are less injurious, and yield enormous crops. There are many wells throughout the district, water existing at moderate depth. The climate of the greater part of the district is more congenial and pleasant to Europeans than that of any other part of India, and the atmosphere is dry and clear. The cold in winter is very severe, descending even sometimes to 10° below zero of Fahrenheit. The hot winds are irregular and unsteady. The rains are limited generally to July and August, when they are heavy. Agriculture succeeds well in Moradabad, the crops being very varied. Maize and millet are the chief sources of food to the lower classes; but almost every variety of esculent vegetation may be grown with little difficulty. The population is very dense, exceeding 400 to a square mile. The chief town is *Moradabad.* It is built on a ridge of ground on the right bank of the Ramgunga. It is the civil station, and had formerly a mint. There is one street about a mile long; but the public buildings are few and insignificant. Distance NW. from Calcutta, 838 miles.

Shajehanpur, including the *Terai Pergunnas*, (lat. 27° 15′—28° 45′; long. 79° 23′—80° 30′; area, 2,446 sq. m.).—These districts are about 800 feet above the sea in the northern extremity, descending to 500 feet on the southern frontier. The descent, however, is gradual. The pergunnas in the north are part of the *Terai*, a tract of marshy forest and jungle, stretching along the foot of the mountains. Innumerable small springs, oozing from the hills, and rising in the ground near their foot, saturate the soil and give growth to gigantic trees, encumbered above with air plants, and below with impenetrable underwood. Grasses and other herbage, attaining a height of ten feet, overrun the more open parts, and are annually set on fire to allow of a more succulent growth, which is fed down for about two months by numerous herds of kine and buffaloes. A malaria broods over the whole region, fatal to man and domesticated animals, but not affecting the elephant, rhinoceros, tiger, panther, leopard, wild buffalo, deer, and other wild animals, which there attain extraordinary size and vigour. This tract gives rise to a large river, and is no doubt capable of drainage and improvement. The southern part of Shajehanpur has a fine climate, little incommoded by hot winds. The town of *Shajehanpur* is on the left bank of the Gurra. It is a large place, with some stately old mosques and a castle, which are ruinous. There is, however, a fair amount of trade. *Powain*, 18 m. NE.; *Tilhur*, 12 m. WNW.; are other towns of small importance.

Non-Regulation Provinces.

Ajmir.—An outlying province of 2,672 sq. miles, surrounded by Rajput states, and in the centre of Rajputana.

Ajmir (lat. 25° 43′—26° 42′; long. 74° 22′—75° 33′; area, 2,029 sq. m.).—The southern part is sandy and level, the northern and western parts hilly, and traversed by ridges of the Aravalli mountains. It is crossed by a few channels, only carrying water in the rainy season, so that water for cultivation depends on the reservoirs or tanks which are scattered in abundance through the district. The soil is impregnated with mineral salts, which injure the water. In the mountains are veins of carbonate of lead, worked from time immemorial. There are also ores of manganese, and indications of copper, besides good iron ore. The population is about 250,000. The chief town (of the same name) is beautifully situated at the base and on the slope of a hill. Some of the streets are wide, and the houses handsome. It contains a large arsenal and military magazine; and in the neighbouring hills is a sanatorium.

Mairwarra (lat. 25° 25′—26° 10′; long. 73° 50′—74° 30′; area, 372 sq. m.).—This very small district is a portion of the Aravalli range, consisting of alternate ridges and valleys, the bottoms of the valleys about 1,600 feet above the sea, and the ridges about 1,000 feet higher. Lead, copper, and antimony ores occur in the hills; cotton, opium, rice, wheat, millet, and barley are the staple products of the valleys. The inhabitants are called Mairs, a savage independent race. The principal place is *Nya Nuggur*, a newly established town.

Jansi.—A small province, consisting of a narrow strip of territory, extending from the Jumna, north of Allahabad division, towards the south-west, into Central India. It was originally a part of the Bundela country, or Bundelcund (*see* p. 166).

Jaloun (lat. 25° 32′—26° 26′; long. 78° 45′—79° 53′; area, 1,544 sq. m.).—One of the districts in Bundelcund now belonging to British India. The town of *Jaloun* was formerly a considerable mart for inland traffic, and is still important. *Calpi* is a town of 19,000 inhabitants, having an extensive trade, and manufacturing paper and sugar-candy. It is one of the hottest places in India.

Jansi (lat. 24° 55′—25° 48′; long. 77° 53′—79° 31′; area, 1,610 sq. m.).—A district in Bundelcund, divided into two parts, separated by a narrow slip of native territory, belonging to the Raja of Tehri. The town of *Jansi* is much frequented by caravans, and has a considerable trade in cloth and carpets. It is also noted for the manufacture of bows, arrows, and spears. On a rock, overlooking the town, is a lofty castellated building of stone, surmounted by a huge round tower. It is a fortress built by the Mahrattas. There is another hill, half a mile south-east of that on which the town is built. Distance N.W. from Calcutta, 740 miles.

Lullutpur (area 1,947 sq. m.) is a district south of Jansi with a small town of the same name.

Kumaon.—An extensive province, including the two districts of Kumaon and Gurwal, having a total area of 11,500 square miles and 600,000 inhabitants. The southern portion (the Babur) is densely wooded, and destitute of streams. Northwards there are rugged mountain masses rising into peaks, many of them above the snow-line. The sides and slopes of some of these mountains are extremely fertile, and many of the valleys are ravines clothed with jungle throughout. In this northern part are many streams, all feeding the Ganges. The general appearance of the province is highly picturesque. Besides the vegetable produce, gold is found in the streams, and both copper and iron ores abound.

Gurwal (*British*). (lat. 30° 2′—31° 20′; long. 77° 55′—79° 20′; area, 5,500 sq. m.).—This district extends over the south-western declivity of the Himalaya, consisting of a lofty mountain range and deep valley. Some of the mountain peaks are upwards of 20,000 feet above the sea: the highest is estimated to exceed 23,000 feet. The climate is hot in the low tracts on the river banks, moderate in the lower regions of mountains, but in the northern part rises above the limits of perpetual congelation. Monkeys are numerous as high as *Suki*, which is 8,869 feet above the sea, and in latitude 31°. The crops in the cooler part of the district include wheat, barley, buckwheat, and other grains, pulse of various sorts, and oil-seeds. The natives are short in stature: their complexions are less dark than those of the Hindus generally, but the hair is invariably dark, with little growth of beard. They are probably a mixed race of Mongol and Indo-European. They inhabit houses substantially built in three stories. Cattle-stalls are on the ground floor, a granary in the middle, and the family live on the upper floor. The roof is nearly flat. They are mild and peaceable, but pilfering is not unknown. The chief town is *Sri Nuggur*, a decayed place on the left bank of the Aluknanda, a feeder of the Ganges. The civil station is *Puri*. At *Badrinath*, 10,000 feet above the sea, is a hot sulphur spring. At *Banassa* are thermal springs.

Kumaon (lat. 29° 5′—31° 6′; long. 78° 17′—80° 56′; area, 6,000 sq. m.).—A mountainous tract, exhibiting extraordinary diversities of elevation, temperature, and climate. The southern part is forest-land or marsh, beyond which is a dry table-land of considerable elevation. The northern part rises into some of the loftiest peaks of the Himalaya, as many as thirty-four summits rising above 18,000 feet. Gold has been obtained from the river sands, but not in large quantities. There are lead and copper ores, but they have not been worked to profit. The climate varies from the suffocating and deadly sultriness of the Terai to the eternal snows of the Himalaya. Snow falls occasionally, but not oftener than about one season in three. Both the botany and zoology of the district are interesting. Leopards and bears are troublesome: the tiger is a great scourge. The *cheang*, an equine quadruped, is seen, but rarely obtained alive. The spotted axis is found in the higher mountain tracts. Large flocks of sheep are bred on the mountains, both for the sake of their wool, and to be employed as beasts of burden. There are many passes over the mountains to Tibet, the traffic over them being carried on by the Botias, who occupy the higher valleys. The people of the province are chiefly Hindus, and their religion is a mixed Braminism and Paganism, almost every peak, forest, rock, and spring having its presiding deity. *Almora*, the chief town, is in the centre of the province, in a beautiful and highly cultivated country, whose climate is said to be like that of Nice. There are quarries of stone and slate in the neighbourhood. The houses of the civil officers are not at Almorah, but at *Hawulbag*, five miles distant. Distance of Almora NW. from Calcutta, 910 miles. *Nyni Tal*, twenty-two miles from Almorah, is the sanatorium of the province. It is situated on the borders of a beautiful lake, 7,000 feet above the sea.

6. Oude.

This great province, stretching from 25° 34' to 29° 6' N. lat., and from 79° 45' to 83° 11' E. long., having a superficial area of 22,456 square miles, occupies the centre of the great sub-Himalaya valley. The Ganges is its southern boundary. Its general surface is a plain, sloping from north-west to south-east, yielding exuberant crops. The Gagra, its principal river (after the Ganges), is navigable for steamboats. The Chowka is its chief tributary. The Gumti, and many feeders, traverse the country from NW. to SE. The climate is dry during part of the year, and there is a rainy season. In summer the thermometer rises to 112°, in winter it sinks below the freezing point. The annual rain-fall varies from thirty to eighty inches. The plains are subject to hot sultry winds from the west, and occasional fierce hurricanes. The soil is exceedingly fertile, the staple products being wheat, barley, maize, bajra, rice, pulse, oil-seed, sugar-cane, indigo, opium, and cotton. Saltpetre and soda are also largely obtained and exported. Teak and other valuable timber trees (local names, "Sal," "Sissu," and "Toon") are common in the forests. The trade of the province is very considerable. Oude is governed by a Chief Commissioner, assisted by four Commissioners, each presiding over five districts. The divisions are Lucknow, Kyrabad, Baiswara, and Fyzabad.

The population is 6,502,884, and consists largely of Hindus, most of whom are Rajputs, a people exhibiting in their appearance, conversation, and habits of life, a proud and martial character, who are accustomed to the use of arms and athletic exercises from infancy, and who prefer military service to every other means of livelihood. They have always furnished the armies of Hindustan with most of their finest men.

Oude was a part of the territory of the sovereigns of Delhi until about 1760, when it was taken possession of by the Viceroy, who had charge of it under the Mogul, and raised into a separate kingdom. After this there was a succession of rulers, more or less under British influence and protection, lasting for nearly a century. In the year 1856, however, after much bad government, and on the refusal of the then king to sign a treaty accepting the direct interference of the British Government, the country was annexed. The king and his relatives were largely pensioned, the fortresses which covered the country were demolished, the people disarmed, and law and order enforced. Roads have since been made, bridges built, police, gaols, hospitals, and schools established, and Oude is now becoming in every respect one of the foremost countries in India. The proposed Lahore and Peshawur railway will cross Oude completely from NW. to SE., following the valley of the Gumti.

The principal town is *Lucknow*, the capital and the seat of government. Population, 300,000. It is situated on the right bank of the Gumti, which is navigable for some distance above the city, and below it as far as the Ganges, and so to the sea. The stream is crossed by two bridges. Lucknow presents a varied, lively, and even brilliant aspect, when viewed from a distance; but is meanly built, the houses being constructed of mud and straw roofs. It is, however, extensive, a continuous mass of buildings extending for about four miles along the bank of the river. Parallel to this, for about a mile, is a wide street, handsome, and with many fine buildings; and between this street (called the Chinka Bazaar) and the river, is the residence formerly occupied by the king. In the north-western part of the city is the celebrated "Imambara," or chief mosque, one of the most elegant and beautiful specimens of the light and fantastic style of Mahomedan architecture in existence. It consists of a series of buildings of great extent. There are many other interesting buildings, among them an observatory: distance from Calcutta, 640 miles. Lucknow is connected with Cawnpore by rail.

Oude, the former capital of the kingdom, is on both banks of the river Gagra, and adjoins the more recent city of Fyzabad. Oude is more interesting for its ruins, which are believed to be of extreme antiquity, than for its present beauty. It enjoys the reputation of having once extended 200 miles in length, and 50 miles in breadth; and it is asserted that Lucknow, 80 miles from the present city, has been one of its suburbs. However this may be, Oude is still a large and populous city, and is much venerated by the Hindus. *Fyzabad*, also called *Bangala*, is in some sense a decayed place, though improving since the British occupation, and now a considerable commercial town, with manufactures of cloth, metal, vessels, and arms. *Roy Bareilly*, fifty-four miles from Lucknow, situated on the river Sai. *Barraitch*, *Pertabgur*, *Gonda*, and *Seetapur*, are other towns of importance.

7. The Punjab.

General Account.—This extensive territory, one of the most important in North-Western India, is so called from two Persian words, signifying "five rivers," in reference to the five great streams that run through it. These are the Indus, the Jelum, the Chenab, the Ravi, and the

Sutlej. The country extends from the River Jumna on the east to the Suliman mountains on the west, and from Cashmere in the north to Rajputana and Sind on the south. (Its limits of latitude are $27\frac{1}{2}°$—35°, and of longitude 69°—79°.) The area is 95,768 square miles. The total population, taken in 1868, is estimated at 17,593,946, very irregularly distributed.

The northern and southern parts of the Punjab differ very greatly in physical features. In the north-east angle is the Alpine region of Kangra. The north-west angle is also mountainous, the entire tract being intersected by mountain ranges, and consisting of a series of valleys encircled by hills. The remainder is plain country, divided naturally into four sections, called *Doabs*,* stretching south-west with a regularity rarely broken by any elevation of importance; but declining imperceptibly from about 1,600 feet above the sea, at Jelum, to 230 feet at Mithun Kote. No country enjoys more largely than the Punjab the means of irrigation and internal navigation, owing to its noble rivers, which, though so far from the sea, are all magnificent streams. The soil consists chiefly of sand and clay, and is almost without stones. The rivers easily and constantly wear away their banks and shift their directions, never pursuing exactly the same course for two years in succession. Towns and villages, therefore, except where limestone and other hard rocks occur, are usually at some distance from the rivers, and cultivation does not extend to their banks. Wherever water is obtainable for irrigation the soil is extremely productive; elsewhere it is altogether

* The Doabs are *Bari Doab*, between the Beas and the Ravi; *Reechna Doab*, between the Ravi and the Chenab; *Jetch Doab*, between the Chenab and the Jelum; *Sind Sagur*, or "Ocean of the Indus," between the Jelum and the Indus. The Bari Doab is the most important, as containing the home of the Sikh nation, and the three great cities of Lahore, Amritsur, and Multan.

arid, or covered with low brushwood, jungle, and reed-grass. This is the case in the centre of each Doab, and affords almost boundless grazing ground for camels, cattle, buffaloes, sheep, and goats. The climate is characterized by much drought, the rain-fall being nowhere great. The rivers are fed by the melting of the snows, and inundate the country in the month of July, which is also the rainy season. The hot weather begins about the middle of April, and the heat is almost intolerable from that time till August, being greater, indeed, than elsewhere in India. Frequent dust-storms then occur, and on calm days spiral columns of dust arise and travel onward, whirling round continually for one or two miles before subsiding. Mirage is then common. In September the heat moderates. October is temperate and agreeable, and from November to April it is cold. Frosts occur at night; but during the day, even at that season, the thermometer rises to 70°.

The most important products are wheat, barley, millet, rice, cotton, hemp, indigo, tobacco, sugar-cane, and pulses. Oil-seeds, melons, cucumbers, and saffron are largely grown, and dates, figs, oranges, mangoes, and other fruits are common. Flax thrives, and tea is grown on the hills. There is little timber, and large trees are very scarce.

Among wild animals, lions, tigers, leopards, panthers, lynxes, martens, stoats, hyænas, wolves, bears, jackals, foxes, otters, wild hogs, porcupines, nilghaus, deer of many kinds, monkeys, and bats, including the vampire bat, may be named. Camels, buffaloes, horses, and sheep are domesticated. Birds include eagles, kites, parrots, pea-fowl, pheasants, partridges, and land and water-fowl. The rivers contain alligators and many kinds of fish. There are many venomous land reptiles. The silk-worm thrives, and bees make excellent honey and wax. Noxious insects are numerous, and the sand-fly in the hot season is worse than the mosquito.

The people include (1) *Sikhs* (*see* note, p. 111). They

are very numerous in some parts, especially about Amritsur and Lahore. (2) *Jats.* These form the bulk of the agriculturists, and are found in all parts east of the Indus. Their chief home is in the Bari Doab. (3) *Patans,* who inhabit the country west of the Indus. (4) *Goojurs,* probably aborigines, and devoted to agriculture. The language commonly spoken by the country people is called Punjabi, and is a branch of the Indo-Germanic family. In the large towns, Hindi is chiefly used, and in the south, near Sind, Hindi is mixed with Sindi. Urdu and Hindi, Persian and English, are taught in the schools. Most of the people (about two-thirds of the whole) are Mahomedans, half the remainder Sikhs, and the rest Hindus. The Sikhs admit no distinctions of caste. The bulk of the people are very poor, and live in wretched mud-built cottages, in villages and small towns. The condition of all classes is steadily improving, though in the remote North-Western Provinces the people are still in a very rude state. Agriculture is the chief occupation, but the manufacturing industry is very considerable and important. Silk and cotton goods are extensively made in most of the large towns, especially in Lahore, Amritsur, Multan, Shujabad, and Leia. The silks of Multan are called *Kais,* and are noted in the Indian markets. Strong cotton cloths are made at Rohun and Hoshiyapur. Carpets, like those of Persia and little inferior to Cashmere, brocades and rich silks, and arms are made at Lahore.

The Punjab is under the control of a Lieutenant-Governor. There are ten provinces, subdivided into thirty-two districts. Over each province is a Commissioner, and each district is placed in charge of a Deputy Commissioner. The administration of justice is supervised by a Judicial Commissioner, and the management of the revenue is in the hands of a Commissioner of Finance.

The education of the Punjab is superintended by inspectors, assisted by native deputy inspectors and several sub-deputy inspectors. There are four educational circles, and over each is an inspector. There is a director over all. There are two colleges and many schools, of different degrees of advancement. In 1864–5, the number of schools was 2,625, with 92,000 pupils. More has been done for female education in the Punjab than elsewhere in India.

The greater part of the Punjab was annexed to the British dominions at the close of the Punjab war in 1849. At that time the country was almost a wilderness; its highways were unsafe, its resources wasted, and its people in misery. The introduction of regular government, effective police, and better education, the suppression of national crimes, greatly improved means of communication, and canals rendering irrigation easy, have already led to increased cultivation and general prosperity.

The means of communication in the Punjab include rivers and canals, ordinary roads and railroads. The rivers are shallow, and the boats therefore made so as to draw but little water. They are usually heavy, broad, sharp at both ends, and carry ten or twelve tons. Down stream they go at the rate of two miles an hour. Against the current they are tracked along the shore, or have a large sheet to take advantage of a favourable breeze. The canals are numerous, but not much used for navigation. First-class roads have been constructed, and railways are now in operation. These are described in pp. 39, 40. There is complete telegraphic communication with all the principal towns.

Canals are in course of construction, which will, when completed, secure a complete system of irrigation for all the doabs or interspaces between the great rivers of the Punjab. Nothing but water is required to secure large crops in all seasons.

There are thirty-four native states in various degrees of feudal subordination to the government of the Punjab, and a treaty of alliance has been made with Affghanistan.

The Lieutenant-Governor is also in constant communication with other states of Central Asia.

Many of the districts of the Punjab are in the Sikh country, or Sirhind.*

Amritsur, or Umritsur.—A small province near Lahore, between the Chenab and the Beas, crossed by the Ravi and several smaller streams. The northern extremity touches Cashmere. Its area is 5,337 sq. miles.

Gurdaspur (lat. 31° 30′—32° 31′; long. 74° 55′—75° 43′; area, 1,341 sq. m.).—A district bounded on the north by Cashmere, and intersected both by rivers and canals. *Gurdaspur* and *Adin Nuggur* are the towns.

Sealcot (lat. 32°—32° 45′; long. 75° 8′—76°; area, 1,960 sq. m.) is to the west of Gurdaspur. The town is on the left bank of the Chenab.

Amritsur, or *Umritsur* (lat. 31° 11′—32° 15′; long. 74° 21′—75° 28′; area, 2,036 sq. m.).—A district bounded by the Rivers Ravi and Beas to the east, and by the provinces of Gurdaspur and Lahore to the north and south. The town of *Amritsur* is a populous city, in the centre of the district. It is a place of pilgrimage, containing a celebrated reservoir, or tank, called "the fountain of immortality." It is fed by springs of very pure water; and on

---

* *Sirhind*, or the Sikh country, is an extensive territorial division of Hindustan, reaching on the north to the Punjab, and bounded on the east by several Hill states and British districts, on the south by Rotuk and Hurriana, and on the west by the state of Bawalpur. It is a vast level plain, 17,000 square miles in extent, rising towards the NE. in the direction of the lower ridges of the Himalaya. It was held till lately by native chieftains. The people are called *Sikhs*, or disciples, and are followers of a religious enthusiast who, in the middle of the fifteenth century, introduced certain modifications of Braminism. Two centuries later there was a subdivision introduced into two great sects, the *Kalisa*, the old Sikhs, and the *Sings*, or Lions. The former more resemble Hindus.

an island in the middle is a richly decorated temple of Vishnu. It has also a large fortress, built by Runjeet Sing. Amritsur has several manufactures of coarse calico and inferior silks and shawls. The streets are narrow; but the houses are built of burnt brick, and are better than those of many districts. A very large transit trade is carried on through Amritsur; and the place is said to be increasing rapidly.

Cis-Sutlej States, or **Umballa.**—A province of about 4,000 sq. m., comprising a group of districts south of the Sutlej river.

Ludiana (lat. 30° 34′—31° 2′; long. 75° 25′—76° 25′; area, 1,359 sq. m.).—A district in Sirhind, on the left bank of the Sutlej, near its confluence with the Beas. The chief town is *Ludiana.* It is built on a bluff rising about thirty feet above a nullah or ravine, parallel to the Sutlej, which it enters fifteen miles below, and of which it was once the channel. It is modern, and has considerable trade. The greater part of the inhabitants are weavers of strong calico. Shawls are also made. Distance NW. from Calcutta, 1,102 miles.

Simla (lat. 31° 6′; long. 77° 14′; area, 18 sq. m.).—A very small district, between seven and nine thousand feet above the sea, in which is a celebrated sanatorium, or retreat for persons whose health is affected by the extreme heat of the plains of India. *Simla* is on a ridge terminated at each end by lofty peaks, and the views from the town are very noble. The snowy summits of the great Himalaya chain are distant about sixty miles, but appear much nearer when their outline is cut out against the clear sky in fine weather. The climate is exceedingly healthy; the thermometer seldom rising above 62° Fahr., except in May, June, and July, when it reaches 72°. These are the rainy months, and during the rest of the year the climate is delightful. Snow falls in January and February, but does not remain on the ground more than two or three days. The people are poor, but simple and tractable, and the practice of polyandry is common among them, one woman having five or six husbands. It is governed by a Deputy Commissioner and there is always a large number of English residents. Distance NW. from Calcutta, 1,097 miles.

Umballa, or *Amballa* (lat. 29° 50′—31° 8′; long. 76° 30′—77° 40′; area, 2,628 sq. m.).—A narrow strip of territory in Sirhind, reaching from the Sutlej to the Jumna, and crossed by a number of streams proceeding from the southernmost and lowest ridge of the Himalaya. It is level and well cultivated, abounding in groves of mango trees, and capable of furnishing abundant supplies. *Umballa*, the chief place in the district, is a large walled town, of very narrow streets, with a fort at the north-east corner. The climate is very hot. Distance NW. from Calcutta, 1,020 miles. *Thunnesir* is built on an irregular mound, formed by the ruins of a former city. The place is surrounded by a ruinous wall, connected with a dilapidated fort. Outside the town is a picturesque Mussulman tomb, also in ruins. At a short distance from the town is the lake of Kurket, about a mile long and half a mile wide. It is sacred in Hindu mythology. Distance NW. from Calcutta, 988 miles.

Delhi.—This important province has an area of about 5,600 square miles. It occupies a long strip of territory on the right bank of the Jumna, between that river and a group of Sikh protected States. It was formerly within the Government of the North-Western Provinces.

Delhi (lat. 28° 24′—28° 54′; long. 76° 49′—77° 29′; area, 1,227 sq. m.).—A small district, with a population of about half a million, more than a third being concentrated in the city of Delhi and its suburbs. The soil is barren and rocky, and remarkable for its saline efflorescence. The wells yield brackish water. The climate is dry and favourable. It is intensely hot in summer, and is exposed to considerable cold in winter. Irrigation is needed for the crops, which are chiefly barley, wheat, and pulse. The city of *Delhi* is the Rome of India in all that relates to the remains of ancient grandeur. The ancient city is said to have extended thirty miles along the banks of the Jumna, and, approaching the modern city from the direction of Agra, the appearance is very striking, from the innumerable ruinous monuments of former prosperity and grandeur. "Everywhere throughout the plain rise shapeless half-ruined obelisks, the relics of massive Patan architecture, their bases buried under heaps of stones bearing a dismal growth of thorny shrubs. Everywhere one treads on overthrown walls. Brick mosaics mark the ground-plan of the humbler dwellings of the poorer classes. Among the relics of a remote age are occasionally to be seen monuments of a light and elegant style of architecture, embellished with brilliant colours, gilt domes, and minarets encased in enamelled tiles. The ancient city was called *Indraprestha*, or *Inderput*, and appears to have existed in the tenth century. At the end of the fourteenth century it was invaded by Tamerlane, and afterwards became the capital of the territory of the potentate popularly called the 'Great Mogul.' The modern city was founded in 1631. It is about seven miles in circumference, and enclosed on three sides by a fine wall, interrupted towards the river. There are large bastions at intervals, a ditch, and a glacis. The number of gates is eleven. The principal street (the Chandney Choke) is broad and handsome; the architecture of the houses varied, and fronted by trees. The other streets are narrower, but clean. The city contains several splendid palaces, of which the imperial palace is about a mile in circuit. It is enclosed by a wall of red granite, forty feet high, flanked with turrets and cupolas, and is entered by a succession of noble and lofty gateways, all of red granite, highly sculptured. The details are superb, and in the best style of Oriental Gothic. The Jumma Musjid, or mosque, is the finest edifice of its kind in Upper India. There are several other fine mosques. Near the city is an enormous observatory. Nine miles distant is the

Kuttub Minar, probably the highest column in the world. It is 242 feet high, and was erected by the Emperor Kuttub, who died A.D. 1210. It tapers regularly from the base to the cupola, which is said to hold a dozen persons." Delhi is 976 miles from Calcutta.

Gurgaon (lat. 27° 40′—28° 30′; long. 76° 21′—77° 35′; area, 2,016 sq. m.).—A low, level, fertile tract on the right bank of the Jumna; the surface of the country being furrowed by deep chasms and ravines, marking the course of the torrents descending to the river. In certain places the ground is saturated with salt, and constantly covered with efflorescent crystals during hot weather; and here the water is brackish, or even salt. In other places, often nearly adjacent, the soil is free from salt and fertile, and the water perfectly fresh. Occasionally fresh water is obtained in shallow wells; but a few feet lower the salt layer is reached. Salt is largely manufactured. The chief town is *Gurgaon*. It is at the foot of a range of hills, and is distant SW. from Delhi 18 miles; NW. from Calcutta 918 miles. *Ferozpur* is a walled town of 10,000 inhabitants, near which iron ore is worked.

Paniput, or *Kurnal* (lat. 28° 50′—29° 48′; long. 76° 40′—77° 16′; area, 2,352 sq. m.).—A level district, intersected by the Delhi Canal. Where not irrigated it is barren, and, like Gurgaon, there are many parts covered with saline incrustations. With irrigation, which is easy, excellent crops are obtained. The town of *Paniput* is the chief place. It is walled; and around it are ruins of ancient buildings and tombs, indicating a much more important place in former times. On the plains around many of the great battles of India have been fought. Distant N. from Delhi 78 miles; NW. from Calcutta 965 miles. *Kurnal* is a large but filthy town, with a handsome mosque.

Derajat.—A province including about 12,500 sq. m. of the alluvial plain that stretches eastwards from the Suliman mountains. Where not irrigated, it is smooth, hard, and clayey, bare of grass, but with bushes and low trees here and there, seldom reaching twenty feet in height. Loose and irreclaimable sand sometimes takes the place of clay. The whole plain is called the *Daman*,* or Border; and the portion of it called the Derajat is so named from the three towns, Dera Ismael Khan, Dera Futti Khan, and Dera Ghazi Khan, which it contains; the word Dera meaning camp, and Futti, Ghazi, and Ismael the names of *Khans*, or chiefs, of marauding tribes. Where duly irrigated, the clay is very productive.

Dera Gazi Khan (lat. 28° 25′—30° 32′; long. 69° 15′—71° 5′; area, 2,319 sq. m.).—The most southerly part of the Derajat. The town of *Dera Gazi Khan* is large, populous, and busy. It is about four miles from the Indus; and not only has an extensive transit trade, but has several manufactures. The 'Gazi,' whose name it bears, flourished about three centuries ago.

Dera Ismael Khan (lat. 30° 31′—32° 22′; long. 69° 40′—71° 15′; area, 7,096 sq. m.).—The continuation northwards of the tract of which Dera Gazi Khan is a part, on the right bank of the Indus. The town of Dera Ismael Khan is ill-built and dull, but of some size, and not without considerable business, especially in spring. The situation is very good for trade, though rather dangerous, as the whole town was carried away a few years ago by an inundation. The modern town is, however, further from the river.

Leia, or *Bunno* (lat. 30° 34′— ; long. 70° 50′—72° 10′; area, 3,150 sq. m.).—A district on the left bank of the Indus, opposite Dera Ismael Khan. The town of *Leia*

---

* The whole district of the Daman is 300 miles long from the Salt range in the north, to the confines of Sind on the south, and has an average breadth of sixty miles. The heat in summer is intense; and the productions are those of the rest of the plains of India. The Daman proper is confined to the tract on the right bank of the Indus, but the division includes land on both sides.

is built on a branch of the Indus about eleven miles east of the river. It is a place of great business, and increasing, being the mart for the abundant and rich produce of the surrounding fertile country. *Cachi* is a subdivision of Leia, and consists of two parts, the Thur, or desert tract, beyond the reach of the Indus, and the low land (called *Cachi*), annually inundated. The former is now arid, and thinly peopled; but the ruins scattered over the country prove it to have been once thickly peopled. *Nurpur*, in the Thur, is a manufacturing town, where blankets are made. Gold, iron, and coal and salt are found in Cachi.

Hissar.—A province including three districts, reaching from the Sutlej nearly to the Jumna, at Delhi, and touching on the north-western boundary the great desert and the province of Rajputana. The total area is about 8,500 sq. m.

Buttiana, or *Sirsa* (lat. 29° 12′—30° 29′; long. 73° 51′ —75° 22′; area, 3,116 sq. m.).—An irregular district, the western part of which, on the skirt of the great sandy desert, is nearly waste and uninhabited. In the middle and eastern parts the soil, when irrigated, is productive. It was formerly less arid, and more cultivable than it is now, and much more thickly peopled. The chief town is *Butnair*, formerly flourishing, but now decayed.

Hurriana, or *Hissar* (lat. 28° 33′—29° 49′; long. 75° 20′—76° 22′; area, 3,540 sq. m.).—A district whose soil is formed of alluvial matter swept down by the Guggur and other sub-Himalaya streams.* It is very fertile when watered, but is dependent on the tanks or reservoirs constructed to collect the periodical rains that fall at the close of summer and beginning of autumn. As the hot season advances these supplies generally fail, and water can only be had from wells, which are deep. Thus for great part of the

* These streams run towards the Indus. They are of considerable magnitude in the rainy season, but at other times are reduced to a thread. They are lost in the sandy desert.

year the country is arid, and want of water an urgent distress. Besides the cultivated part, there are jungles and waste, tenanted by wild animals. The population is not large. *Hissar* is the chief town, and a great stud depôt. It is on a branch of the Delhi Canal, and has been prosperous, but is now of little importance. *Hansi* is also on this watercourse. Neither place is of much importance. Distance of Hansi NW. from Delhi, 89 miles; from Calcutta, 989 miles. Near Hansi is a great salt mart, where salt of all kinds is manufactured.

Rotuk (lat. 28° 38′—29° 16′; long. 76° 10′—77° 4′; area, 1,823 sq. m.).—A district near Delhi, formerly under the Lieutenant-Governor of the North-West Provinces, but included in the Punjab government. Its chief town is *Rotuk*. It has a large population, and a good bazaar.

Lahore.—An important province of about 9,000 square miles, extending from the Chenab to the Cis-Sutlej district of Ludiana. It includes three districts, and the important city of Lahore, the capital of the Punjab.

Ferozpur (lat. 30° 10′—31° 15′; long. 74°—75° 25′; area, 2,692 sq. m.).—This district is in Sirhind (*see* note, p. 111), and was formerly a place of some importance. It is thinly peopled, the greater part being barren, or covered with jungle; but, in former times, it must have been more flourishing. The climate is not unfavourable. The chief town is *Ferozpur*, and has been large, as it stands in the midst of extensive ruins. There is an irregular fort, and the town is defended by a ditch and mud wall. The town and fort have been much improved lately, but the houses are chiefly of mud, and the streets narrow and dirty. *Sobraon*, a village on the left bank of the Sutlej, near which an obstinate battle was fought, in 1846, between the British army, of 15,000 men, under Sir Hugh Gough, afterwards Lord Gough, and a Sikh force of 30,000 men, formidably entrenched. The Sikhs were defeated with great slaughter. *Mudki*, a small town, is also remarkable for a victory over the Sikhs in the war of 1845. The number of English officers returned as killed or wounded exceeded fifty.

Gujranwala, or *Gujuruwalla* (lat. 31° 35′—32° 30′; long. 73° 30′—74° 30′; area, 2,657 sq. m.).—A district in the doab or interspace between the Ravi and the Chenab rivers. It is traversed by another stream. The chief town was the birth-place of the ancestors of the celebrated Runjeet Sing, whose grandfather was a common soldier. There is a fort, the interior of which is highly decorated.

Lahore (lat. 30° 40′—31° 55′; long. 74°—75°; area, 3,624 sq. m.).—This district is intersected by the Ravi river, the greater part of it lying in the doab between the Ravi and the Sutlej, and the rest between the Ravi and the Chenab. The former is poorly cultivated and thinly inhabited. The chief place is *Lahore*, the capital of the Punjab, and the seat of local government. It is a large city, the circuit of the line of fortifications exceeding seven miles, and is surrounded by a brick wall, formerly twenty-five feet high. The fort or citadel contains extensive magazines. The interior of the town is heavy, the streets being narrow and dirty, and many of the houses lofty and gloomy, enclosed within extensive dead walls. There are, however, several large and handsome mosques, which for the most part have been desecrated by the Sikhs. Among them, the Padsha mosque is the most remarkable. It is massive and lofty, built of red sandstone, and ornamented with spacious cupolas. The tomb of the Mogul Emperor Jehangir is a very extensive and beautiful structure, also of red sandstone, and richly decorated. It is three miles west of the town, on the other side of the Ravi. About the same distance north-east of the town is the garden of Shah Jehan, a superb monument of Eastern magnificence. It is about half a mile long, with three successive terraces rising one above the other, and contains 450 fountains, which cool the air. The water is subsequently collected into marble reservoirs. Near Lahore is *Mian Mir*, a large military cantonment. The population of Lahore is about 100,000. Distance NW. from Calcutta, 1,156 miles.

Multan.—This province includes the interspace or doab between the Punjnud and Sutlej on the east, and the Indus on the west, being terminated northwards by the district of Leia on the west, and Gujranwalla on the east. The total area is about 20,000 square miles.

Gogayra, or *Montgomery* (lat. 30°—31° 10′; long. 72° 21—74° 40′; area, 5,577 sq. m.).—A level and low district, formerly crossed by the main channel of the Sutlej, which now runs several miles to the east. The town of *Gogayra* is on the Ravi, which bounds the district to the west.

Jung (lat. 30° 35′—36° 52′; long. 71° 50′—73° 48′; area, 5,712 sq. m.).—This district occupies a considerable part of the Rechna Doab, between the rivers Chenab and Ravi, two of the main feeders of the Indus. It is a wild, bare tract of land, with the exception of the strips of land along the banks of the rivers.

Mozuffurgur (lat. 28° 55′—30° 43′; long. 70° 30′—71° 52′; area, 3,022 sq. m.).—A long narrow strip of land, or doab, between the Indus on the west and the Punjnud and the lower part of the Chenab on the east. The chief place is *Mozuffurgur*. It is on the right bank of the Chenab, a little below Multan.

Multan (lat. 29° 23′—30° 35′; long. 71° 21′—73°; area, 5,882 sq. m.).— A large and important district in the doab between the Chenab and the Sutlej, reaching to the confluence of the streams. The part subject to the inundations of the Chenab produces rich crops. The city of *Multan* is one of the principal places in the Punjab. It is built on a high mound of ruins, consisting of the remains of older towns, on the same site; and the neighbourhood is covered with fragments of buildings, some of them of great extent and beauty. The population is now 80,000. The bazaars are extensive and well supplied, and the merchants are rich. Its manufactures are silks, cottons, shawls, brocades, tissues, &c.

Peshawur.—This province includes three districts, occupying the north-western corner of the Punjab, watered by the Indus, and crossed by the Cabul branch of that river. Cabul is its western frontier, the higher part of the Himalaya range its northern, and Cashmere its eastern frontier; and it reaches Affghanistan, near the commencement of the celebrated Kyber Pass. Its total area is nearly 8,000 square miles.

Hazara, or *Huzara* (lat. 33° 42′—35° 1′; long. 72° 35′—74° 10′; area, 3,000 sq. m.).—This district consists of a long narrow strip of land of crescent shape, bounded on the east by the Cashmere frontier, and on the west by a mountain range separating the Upper Jelum from the Indus. The Indus itself forms the south-western boundary. It is very mountainous in some parts. *Huripur*, or *Haripur*, is a populous and thriving place, on the river Dor, which falls into the Indus about ten miles beyond. Its bazaars are very well supplied.

Kohat (lat. 32° 13′—33° 50′; long. 70° 23′—72° 20′; area, 2,838 sq. m.).—A district south of Peshawur, on the right bank of the Indus, in the hilly tract north of the Salt Range, and including the valley of Kohat. It is populous, fertile, and well watered by the River Teo. Its beautiful situation, and the luxuriant vegetation of the surrounding country, render it a delightful place. The town of *Kohat* is meanly built, but has a good bazaar and fine mosque. At *Sikh*, a few miles east of the town, are springs of naphtha and deposits of sulphur.

Peshawur (lat. 33° 42′—34° 30′; long. 71° 35′—72° 42′; area, 1,929 sq. m.).—The extreme north-western corner of British India, lying between the Indus and the Kyber mountains. The climate is very hot in summer, but the heat is mitigated by the cool breezes from the mountains. The country is naturally fertile, and is well watered. The crops are wheat, barley, maize, millet, and esculent vegetables. Some of the best rice in the world is grown in this district. It is called Bara rice. Melons of all kinds, cucumbers, pumpkins, and gourds, are grown in great abundance. The town of *Peshawur* is the capital of the district, and is only twelve miles from the opening of the Kyber Pass. It is situated in a rich and fertile plain, watered by the Cabul river. The houses are built of mud or unburnt bricks, and the whole city has a melancholy appearance, presenting numerous ruins of great dimensions, the result of recent violence. There are many fine mosques, but they have been polluted by the Sikhs, and are going to ruin. This city was the ancient capital of Eastern Afghanistan; it is now one of the largest military stations under the government of British India. *Attock* is a picturesque and formidable-looking fortress on the left bank of the Indus. *Kyrabad* is on the opposite or Peshawur side. Both towns are small and much decayed.

Rawul Pindi.—A province of nearly 17,000 square miles, between the Upper Indus and the Chenab, traversed by the Jelum, bounded by Cashmere on the north-east, and the Peshawur province on the north-west. The range of the Salt mountains crosses the territory.

Gujerat (lat. 32° 10′—33°; long. 73° 25′—74° 41′; area, 1,785 sq. m.).—A district in the upper part of the doab between the Jelum and the Chenab, touching the Cashmere territory in its north-eastern borders. It is watered by some tributaries of the Chenab. The capital (of the same name) is a walled town of considerable size. Near it a great battle was fought in 1849, between a force of 25,000 British, under Lord Gough, and a Sikh army of 60,000 men. The Sikhs were defeated, and lost fifty-three pieces of artillery.

Jelum (lat. 32°—33°; long. 71° 20′—73° 55′; area, 3,910 sq. m.).—A remarkable district, crossed by the mountain chain called the Salt Range, stretching from the eastern base of the Suliman mountains to the River Jelum, which gives its name both to the province and its capital. The Salt Range is so called from the number and thickness of beds of common salt which it contains in many places. The salt mines have been long worked, and in 1832 yielded at the rate of 40,000 tons per annum. Salt, alum, antimony, and sulphur are obtained.

These mountains are not lofty, nowhere exceeding 3,000 feet. The soil in and near the range is barren, vegetation is scanty, and the bold and bare precipices present an appearance of great desolation. The town of *Jelum* is in the extreme north-eastern corner of the district, on the right bank of the river. It is of considerable extent, and upwards of 1,600 feet above the sea. The population is chiefly Mahomedan. The climate is unhealthy, owing to the inundation which spreads on the eastern bank of the river. *Jelalpur* is a town on the Jelum below the chief town, and is built near one of the great passages across that stream, on the route from Hindostan to Afghanistan. Either Jelalpur or Jelum was the scene of the great battle between Alexander the Great and Porus. *Rotas*, near Jelum, is a strong fortress, celebrated in the early history of the Mahomedans in India as one of their main bulwarks between Tartary and Hindostan.

Rawul Pindi (lat. 33°—34°; long. 71° 40′—73° 50′; area, 6,216 sq. m.).—A district immediately north of the district of Jelum, between the Indus and the Cashmere frontier. The chief town is large and populous, with a large bazaar. It is surrounded by a wall with bastions, and has an old fort. Population about 16,000. Between Rawul Pindi and Peshawur is *Hoosan Abdul*, now a paltry place, but celebrated in Hindu mythology. Here is preserved the hand of Nanac, the founder of the Sikh religion. Among the ruins is the tomb of Husan Abdul.

Shahpur (lat. 31° 40′—32° 40′; long. 72° 18′—73° 23′; area, 4,698 sq. m.).—A district in the Jetch Doab, between the Jelum and the Chenab. The chief town is *Shahpur*, on the left bank of the Jelum.

Trans-Sutlej States.—A province situated between the Sutlej and its main feeder, the Beas, extending from the confluence of those streams towards the north, and reaching the lower ranges of the Himalaya, where several of the feeders of the Beas take their rise. It includes three districts. Area, 6,245 square miles.

Hoshiyapur (lat. 31°—32° 6′; long. 75° 35′—76° 40′; area, 2,086 sq. m.).—A strip of country having a low range of mountains running along its whole length, from which proceed several streams, some running into the Beas and some into the Sutlej. The town is small, and bears the same name.

Jalundur (lat. 30° 37′—32° 39′; long. 75° 3′—76° 8′; area, 1,333 sq. m.).—A fertile tract in the doab bearing the same name. It has an agreeable climate. The town also, *Jalundur*, is ancient, and surrounded by the remains of former greatness. It is a military station.

Kangra (lat. 31° 55′—32° 23′; long. 76° 12′—76° 43′; area 2,826 sq. m.).—A hilly district, favourable for the cultivation of tea, lying among the mountains in the lower ranges to the south of the great Himalaya chain. It possesses a hill-fort of considerable interest and great strength (*Kot Kangra*), constructed on the top of an eminence about 150 feet above a small stream near the Beas river. The eminence is about three miles in circuit, bounded almost all round by inaccessible precipices, and strengthened by masonry and ramparts. The fort was defended for four years against the Goorkas, but was finally taken by Runjeet Sing. Its position is exceedingly strong.

8. Native States of North-Western India.

The number of the native states still remaining in North-Western India, all more or less under British influence, but ruled by native rajas, is exceedingly large. A few are extensive and important, some, as Rajputana, being extensive countries comprising groups of states; but many are very small indeed. The following list contains the names of all the states:—

Bahadurgur (area, 48 sq. m.; pop. 14,400).—A small state adjacent to the British district of Delhi. It has a considerable walled town of the same name.

Bawalpur (lat. 27° 41′ — 30° 25′; long. 69° 30′ — 73° 58′; area, 25,200 sq. m.; pop., probably 925,000).—A feudatory state under British management. It consists of a narrow tract 310 miles long, on the left bank of the Sutlej and Indus, reaching eastwards to the great desert of Rajputana. It is a remarkably level country, having no eminences beyond sand-hills fifty or sixty feet high. The only cultivable part extends along the river line for a distance of about ten miles from the course of the stream. Beyond this all is loose sand. The cultivated part yields luxuriant crops of wheat, rice, tobacco, and indigo, and groves of trees. It is inhabited by a mixture of Hindus, Beluchees, and Affgans. The higher classes speak Persian, the lower a dialect of Hindi. It is governed by a ruler styled the Khan. The chief town is *Bawalpur*, a mean collection of houses of unburnt brick, surrounded by a ruinous wall of mud. It is on the Sutlej, about midway between the two extremities of the country. It is a place of some trade, and has silk manufactures; population, 20,000. *Ootch*, near the confluence of the Jelum, is a large town in the midst of a fertile country. *Kyrpur* is a town on the left bank of the Gara, with a tolerable bazaar and some trade, in a part of the sandy waste constantly and rapidly encroached on by blown sand. It is the only place between the banks of the Indus and the desert. At *Dirawul* is a strong fortress, difficult of access.

Bullubgur (area, 190 sq. m.; pop., 57,000).—A small district adjoining the British district of Delhi.

Cashmere, or **Kashmir.** (Ranir Sing's dominions.) (Lat. 32° 17′—36°; long. 73° 20′—79° 40′; area, 60,000 sq. m.; pop., about 3,000,000.)—A name now given by the English to an extensive tract of country, reaching from the

Punjab on the south and west, to Tibet on the north and east, its extreme length being 350 miles, and its breadth 270. It includes the valley of Cashmere, or Cashmere proper, and several provinces of smaller importance, almost entirely mountainous. In the southern portion, the mountain sides are clothed with forests of cedar and pine; but further northwards, towards Tibet, there are large tracts destitute even of a trace of vegetation. The scenery of the mountains is in the highest degree picturesque. "The grandeur and splendour of Cashmerian scenery results from the sublimity of the huge enclosing mountains, the beauty of the various gorges, the numerous lakes and fine streams, rendered often more striking by cataracts, the luxuriance and variety of the forest trees, and the rich and varied vegetation of the lower ground."

The valley of Cashmere is an expansion of part of the valley of the Jelum, here called the Behut. It is an oval plain, 4,500 square miles in area, surrounded on all sides by lofty mountains, whose passes on the north side are not less than 10,000 feet above the sea. The actual plain, or bottom of the valley, is about fifty miles long and ten to twelve miles wide, and is 5,300 feet above the sea. The river flows through it in a tranquil navigable stream, winding about, at one time washing the base of the hills on one side, and then crossing to the other. On the rising of the river in summer, when the snow melts on the mountains, the whole plain would be inundated but for a system of dykes called *bunds*, placed along the course of the stream. The river expands into lakes, one of which is twenty miles long by nine broad, and another as long, but only half the breadth. There are also swamps. The soil is very fertile, and produces all sorts of corn and fruit, and flowers, especially roses, which are cultivated for distillation. Among the natural productions is a nut, the seed of a water-plant growing in the larger lake. Many thousand tons of it are annually collected and used as food.

The inhabitants of Cashmere exhibit a mixture of the

Hindu and the Tartar, the Tartar characteristics being more marked as we approach the mountains; the dress, customs, and even the religion, changing gradually with the changes of the physical features of the country. The inhabitants of the valley are chiefly Mahomedans of the orthodox or Sunni class. The people are tall, well formed, and intelligent; their language, called *Cashmiri*, is derived from Sanscrit and Persian; their dress is chiefly woollen. Their houses are built of wood and brick, or stone.

The climate of the country is divided into the four seasons, as in Europe, as the periodical rains of India do not reach so far into the mountains. Spring and summer are unhealthy, March and April are rainy, May and June dry and fine, July and August are marked by thunderstorms. Winter lasts four months, and the ground is then covered with snow.

The most celebrated manufacture of Cashmere is that of shawls, which are made of two kinds of wool; one from the tame goat, the other from the fleece, not only of the wild goat, but of the yak, wild sheep, and other animals, even including the dog. The wool used is a fine down, growing close to the skin under the common coat. Atta of roses is made in large quantity and of the finest quality in Cashmere, the roses being especially cultivated for that purpose. Fire-arms, saddlery, leather, lacquered ware, and even paper, are largely manufactured.

Cashmere, or *Sri-Nuggur*, is the capital. It is a large town on the banks of the Jelum. The streets are narrow, and the houses built with a wooden framework as foundation, a stone story above, and a brick upper story. The river is crossed by wooden bridges. *Iskardo*, on the Upper Indus, has a very strong fort. *Islamabad* is a town of some importance, and is distant from the capital thirty-five miles. There are several smaller towns, of which *Kishtawar*, on the Chenab, and *Leh*, on the Indus, are the principal.

The whole territory is under the government of a native Raja, placed in possession in 1846 by the British, to whom the country had been ceded in 1845 by the Sikhs. It had been in their possession since 1819, when they took it from the Affgans, who had conquered it in 1752. The ruler of Cashmere is bound to furnish troops when required.

Deojana (area, 71 sq. m.; pop., 6,390.)—A small state or jaghire in the neighbourhood of Delhi.

Furrucknuggur (area, 22 sq. m.; pop., 4,400).—A small chieftainship of a few villages near Delhi.

Gurwal (lat. 30° 2′—31° 20′; long. 77° 55′—79° 20′; area, 4,500 sq. m.; pop., 200,000).—This state is within the territory governed by the Lieutenant-Governor of the North-Western Provinces. It consists of a group of mountain peaks of great elevation, intersected by numerous valleys draining into the Ganges. Several of the peaks are more than 20,000 feet above the sea. There are no less than twelve streams proceeding from the high ground in this little state. The climate is hot in the low tracts on the banks of the rivers, while the upper districts rise above the limit of perpetual congelation. The natural productions vary exceedingly, according to the temperature; for, while sugar-cane and cotton, rice, ginger, turmeric, yams, and hemp are grown in the low grounds, wheat, barley, buckwheat, and pulse are cultivated in terraces on the mountain slopes, and walnut and other trees grow in the forests at a higher elevation. The natives are of short stature, mild and peaceable, and are of mixed Mongol and Hindu origin. A portion of the country was separated from the rest in 1815, and now forms what is called British Gurwal, in the province of Kumaon (*see* p. 102). The remainder is governed by a native Raja. Gurwal contains some of the sources both of the Ganges and the Jumna. *Tiri*, the chief town, is little more than a village. It is the residence of the Raja.

Hill States.—Under this name are included a number of small independent states occupying a position between the upper Ganges and the Sutlej river. They lie to the north and east of the district of Umballa, in the Punjab, and all, with the exception of three, are on the eastern side of the Sutlej (*Cis-Sutlej*). They are highly picturesque and rugged. There are hardly any other roads through them than sheep and goat paths; their trade is very small, and their manufactures confined to coarse woollen cloths and iron. The people are small in stature, hardy, superstitious, and ignorant. They live in houses three stories high, having a wooden foundation and stone superstructure. They are governed by chiefs, who are tributary to the British authorities. The total area of the group, twenty-eight in number, is nearly 10,000 square miles, the population being 733,500. Those on the western side of the Sutlej are much the largest. The following are the states:—

Cis-Sutlej Group.

Bagul (area, 100 sq. m.; pop. 22,305).—A mountainous district, from which rise several lofty peaks.

Beeja (pop. 981).

Belaspur (*Kulore*) (area, 150 sq. m.; pop. 64,848).—A small state at a high elevation, having three towns and ninety villages. The town of *Belaspur*, the principal place, is flourishing and well built. It adjoins Simla.

Bussahir (lat. 30° 56′—32° 8′; long. 77° 34′—78° 52′; area, 3,000 sq. m.).—This state is intersected by the Sutlej, which divides it into two parts, the northern called *Kunawur*, and the southern *Bussahir*. In Kunawur are extensive and rich deposits of copper ore; and iron ores are very abundant, and much worked. They are of the kind called "specular iron." The whole country is mountainous, most parts being from 7,000 to 12,000 feet above the sea. In Kunawur very fine grapes are produced in great abundance. Of the human population, the upper classes are Rajputs, the other classes Bramins, Cunnoits, and Coolies; but, although of Hindu origin, the observances of Braminism are very imperfect, except that beef is never eaten. There are several poor towns.

Buji, or *Biji* (area, 70 sq. m.; pop. 9,000).—A narrow strip of land on the left bank of the Sutlej, adjoining the Punjab.

Bulsun (area, 64 sq. m.; pop. 4,892).—A small state between the Sutlej and the Tons.

Dami (area, 25 sq. m.; pop. 2,853).—A little state 4,000 feet above the sea, crossed by a tributary of the Sutlej.

Durkothi (area, 5 sq. m.; pop. 612).—The smallest of the states. It contains the peak of Tungru, 10,000 feet above the sea.

Gond (pop. 963).—A small state north of Bulsun.

Hindur, or *Nalagur* (area, 233 sq. m.; pop. 49,678).—A state of some importance, traversed by a steep range of hills, but having a populous and well cultivated valley. The country is picturesque and fertile.

Jubul (area, 330 sq. m.; pop. 17,262).—It is of high elevation, and part of it lies in the beautiful and extensive valley of the Pabur.

Kothar (pop. 3,990).

Koti (pop. 3,981).—A few ridges and intervening valleys.

Kumhassin (pop. 7,829).

Kunea (area, 12 sq. m.; pop. 1,906).

Kunthal (pop. 18,083).

Mangol (pop. 917).

Milog (pop. 7,358).

Mudan (pop. 2,431).

*Puttiala** (pop. 48,836).

Saugri (pop. 1,994).

*Simla** (pop. 31,858).

* The two Hill states of Puttiala and Simla are portions of other districts so named. The rest of Puttiala is a Sikh state (*see* p. 138). Simla is in the Cis-Sutlej states province (p. 112).

Sirmor (area, 1,075 sq. m.; pop. 75,595).—A hilly state, almost the whole of it within the drainage of the Jumna. The chief elevations are on the northern frontier, and are about 12,000 feet above the sea. From these the country falls generally (though broken by hills) towards the south and south-east. The northern extremity has very little rain; but very large and excellent cròps are everywhere to be obtained by irrigation. The state is governed by a Raja, who resides at *Nahun*, the only town of importance, and described as being cleaner and handsomer than the generality of native Indian cities.

Theog (pop. 4,423).—A small territory with a fort, between Simla and Kotgur, 8,000 feet above the sea.

Turre (pop. 3,082).

Trans-Sutlej Group.

Chamba (area, 3,210 sq. m.; pop. 120,000).—A feudatory state under British management, situated and crossed by the Ravi and Chenab. It touches the frontier of Cashmere. The chief town is picturesquely situated at the foot of one of the lofty snow-covered peaks of the Himalaya. It was formerly a place of some trade, and has very greatly improved since the Raja was assisted by a British officer, deputed for that purpose, since 1863.

Mandi (area, 1,080 sq. m.; pop. 139,259).—A small territory on the northern slope of the Himalaya, comprising a number of ridges with the valleys between them. The town of *Mandi* is at the confluence of the Sukyt river with the Beas, which quite intersects the territory. The rivers coming down from the snowy mountains are subject, during summer, to a diurnal rise and fall. Iron and salt are obtained from the mountains.

Sukyt (area, 420 sq. m.; pop. 44,552).—A state on the north side of the Sutlej, with a population of about 45,000. There is a small town of the same name. The state is governed by a Raja.

Jujur (lat. 27° 55′—28° 35′; long. 75° 55′—76° 58′; area, 1,230 sq. m.; pop. 110,700).—An important and well managed state, crossed in two directions by high roads, with its chief town only thirty-five miles from Delhi. It has several streams, which discharge into the Jumna close to Delhi. There are three towns besides the capital.

Kuppurchulla (area, 598 sq. m.; pop. 212,721).

Loharu (area, 200 sq. m.; pop. 18,000).

Patowdi (area, 74 sq. m.; pop. 6,660).—This and the two last-named states are near Delhi. There are towns in each.

Rajputana (lat. 23° 35′—29° 57′; long. 70° 5′—77° 40′; area, 120,263 sq. m.; pop. 9,375,000).—An immense tract of country, extending from Sind on the west to the province of Agra on the east, skirting the Bombay Presidency on the south, and stretching to the Punjab on the north. It includes the region of the Thur, or Great Desert, and the Aravalli mountains. It comprises eighteen independent states, besides the small British province of Ajmir, almost in the centre of the district (*see* p. 100). The following are the native states:—

	Area in square miles.	Population.
Banswarra	1,500	150,000
Bikanir	17,676	539,000
Bundi	2,291	220,000
Burtpur	1,974	650,000
Dolepur	1,626	500,000
Dungurpur	1,000	100,000
Jallawar	2,500	220,000
Jessulmir	12,252	73,700
Jeypur	15,000	1,900,000
Jodpur, or Marwa	35,672	1,783,000
Kerowli	1,878	188,000
Kishengur	720	70,000

	Area in square miles.	Population.
Kota	5,000	433,000
Oudeypur, or Meywa	11,614	1,161,400
Pertabgur	1,460	150,000
Serohi	3,000	55,000
Tonk	1,800	182,000
Ulwur	3,300	1,000,000

Each of these states is governed by a Raja, except Tonk, which is under a Mussulman Nawab, and Burtpur and Ulwur, which are Bat principalities. The whole group is under the political superintendence of an agent, appointed by the Viceroy, to whom each state sends a *Vakil*.

Though so large a country, Rajputana is one of the least interesting in India, much of it being a desert of moving sand, destitute of vegetation and water, and therefore of inhabitants. The western states, Bikanir, Jessulmir, and Marwa are almost desert; but rocky hills appear above the sand in Marwa; and there are said to be valuable marbles and building stones in that state. There is also a fair population pursuing some industries, especially in the districts watered by the Aravalli, one of the few rivers crossing the territory. The tracts on the eastern side, towards the North-Western Provinces, are fertile and highly cultivated. This is the case especially in Burtpur and other states through which rivers pass. The Aravalli range crosses the south-eastern states. The chief streams that water Rajputana are the Chumbul, the Luni, and the Bunass; but for the most part water is only obtained from wells. These are shallow in the eastern states; but in the west, two or three hundred feet deep. There are salt lakes and brine springs in the desert, from which much salt is made. The crops in the east and on the river banks are corn, cotton, sugar, tobacco, and opium. Where there is pasturage, there are large herds of camels, horses, and sheep. In the desert, and in the southern part, are wild asses, nilgaus, and antelopes, besides lions and leopards, tigers, wolves, hyænas, jackals, and foxes. The wild ass is a very fine animal, and is even used for food. He is never seen in bad condition.

The people of this extensive country include several races. The greater number, and the highest class, are Rajputs, a tall, vigorous race, athletic, and of very warlike habits. They claim to be descended from a warrior caste, whose ultimate parents are the sun and moon. The feudal system prevails among them, and the chiefs have great influence. There are several classes recognized among them; one of these consists of a tribe called Batties, who inhabit the western states. They are much addicted to the use of opium, and, when not under its influence, are said to be little better than idiots. They are a dissipated race. Another class is called Bats, who are bards, and sing the praises of their own tribe, and satirical songs concerning their rivals. The Charuns are the priests and historians, and possess great influence, owing to a superstitious notion among the rest of the people that ruin will attach to any one who sheds their blood. They accompany travellers, to protect them from robbers. The Jats are also numerous. The rest of the people are Bramins and Jains.

The chief manufactures of the people who inhabit Rajputana are cotton and woollen goods, carvings in ivory, and working in metals: all these handicrafts are chiefly carried on in the eastern states.

There are few good roads in Rajputana, and neither canals nor railroads. One good road crosses the country westwards from Agra, as far as Ajmir, and then turns southwards into Central India. The telegraph is also carried to Ajmir, and not beyond. The climate and absence of cultivable soil have much to do with this; but the divided interests of the Rajas of the separate states much more; and it must be long before the country is sufficiently advanced to take rank among the important districts of India.

The following notices of the towns of Rajputana include all that possess any interest:—

Abu, in Serohi, is the residence of the Government Agent. It is a sanatorium in the Aravalli range, and is situated near the highest summit of that range, which is about 5,000 feet above the sea. It is a celebrated place of pilgrimage, and possesses one of the most superb of all the places of worship in India, said to have cost eighteen millions sterling in building, besides half a million in levelling the ground for a site. There is a group of four marble temples at *Dilwara*, about midway up the mountain, and five miles from the highest summit, called Mount Abu. They are about four centuries old, and are dedicated to the Jain worship, intermediate between that of Brahma and Buddha.

Banswarra is a large enclosed space, much of it occupied by gardens, and with a handsome palace, some good Hindu temples, and an extensive bazaar.

Bikanir has the appearance of a great and magnificent city, and has some fine houses and temples; but most of the houses are mud hovels painted red, and it stands in the midst of a desolate plain.

Bundi.—A walled town without much trade, with two good bazaars, and a very fine palace adjacent. The town is well built, and is in the midst of a beautiful country.

Burtpur.—A large modern town, with narrow, dirty, crowded streets, and houses built of stone. It was once strongly fortified, and has been twice besieged. Its fortifications are now destroyed.

Dolepur (near the Chumbul).—An ancient place, with some fine antique mosques and mausoleums of the fine freestone of the country, worked into beautiful trellis-work, which is perfectly preserved, though about two centuries old.

Dungurpur.—A fortified town of considerable size.

Jessulmir.—A singular and interesting town, of 20,000 inhabitants, in a rocky tract of yellow limestone. It is fortified by ramparts and bastions of uncemented stone.

Within the ramparts is a citadel on an eminence, three-quarters of a mile in circumference, with steep sides scarped all round. The palace within the citadel is surmounted with a huge umbrella of metal, supported on a stone shaft. There are also six temples in the citadel. The houses in the town are well built. The town is said to have been founded in the twelfth century.

Jeypur.—A large, walled, modern city, situated in a natural basin, the ancient bed of a lake, and surrounded by barren stony hills. On a hill behind the town is the citadel. The town is two miles long and a mile wide, and is one of the best built cities in India; the houses are finely constructed of stone, and the streets are at right angles to one another. The palace occupies the centre, and is half a mile long. The garden is very beautiful; and there is an observatory. This is the residence of a Political Agent of the British Government.

Jodpur, or *Marwa.*—The capital of the most extensive of the Rajput states, governed by a Maharaja. A Political Agent of the British Government resides there. It is situated in a cultivated but woody plain at the southern extremity of a ridge of rock twenty-five miles in length, two or three miles broad, and between 300 and 400 feet above the surrounding plain. It is enclosed by a rampart five miles in circuit, and the general effect from a distance is superb. The streets, however, are irregular and badly laid out, though there are many very handsome edifices of stone. There are several reservoirs (tanks) within the walls. Near the town are fine gardens, and a beautiful building called the Pearl Palace.

Kerowlee.—A large walled town in a cultivated district, difficult of access. The houses are well built, but the streets narrow and filthy.

Kishengur.—A town once considerable but now ruinous, built among hills of granite, and surrounded by a thick wall of masonry.

Kota.—A large and well-built town on the Chumbul, enclosed by a strong rampart, and enclosing a very beautiful palace. It is a thriving and wealthy place, and has a resident Political Agent of the British Government.

Machery.—A small town, formerly the capital of the state of Ulwur.

Oudeypur, or *Meywa.*—The capital of one of the largest states bearing the same name. It is pleasing in appearance, but ill built and unhealthy. The palace is a noble pile of granite, a hundred feet high, with a beautiful artificial lake formed by an embankment thirty-seven feet high, faced with marble.

Pertabgur is a large town, but contains nothing remarkable.

Serohi.—A town built in the fifteenth century upon the ruins of a much more ancient city, still observable. It is a large town, and has much trade. It is celebrated for the excellence of the sword-blades made there.

Tonk.—A large walled town with a mud fort, embellished with several public buildings.

Ulwur, capital of the state of that name, is an ill-built town, surrounded by a wretched mud wall. It contains the palace of the Raja, and an elegant pavilion of white marble.

Rampur (lat. 28° 30′—29° 11′; long. 78° 55′—79° 30′; area, 1,140 sq. m.; pop. 390,232).—A state within the British province of Rohilcund. It is a level, fertile country, abundantly supplied with water in the northern part by two streams, and crossed in the southern part by the Ramgunga. The northern part adjoins the "Terai," and is terribly unhealthy. The town of *Rampur,* on the Kosilla, a tributary of the Ramgunga, is large and irregularly built, but is beautifully situated in a cultivated district. It has a lofty mosque in the market-place. Distance NW. from Calcutta, 789 miles.

Shapura (area about 1,500 sq. miles).

Sikh States.—A group of small native states under British protection, between the Simla district and Hissar province of the Punjab. Their inhabitants were formerly called Malwa Sikhs. Area, 7,070 sq. m.

Furid Kote (lat. 30° 40′—30° 56′; long. 74° 22′—75° 9′; area, 604 sq. m.; population about 75,500).—A small state, almost surrounded by the British district of Ferozpur. The town bears the same name.

Jind.—A territory comprising a number of detached portions, having a total area of 683 square miles, and a population of about 162,920. It is chiefly within or on the northern border of the state of Puttiala. *Jind* is a considerable town, with a good bazaar and palace, but the country around is much overrun with jungle.

Kulsea (area, 155 sq. m.; pop. 62,000).

Molair Kotli (area, 165 sq. m.; pop. 46,200).

Mumdote (area, 370 sq. m.; pop. 37,100).

Narba, or *Nabha* (area, 658 sq. m.; pop. 184,240).—A territory consisting of detached portions. The greater part is a long very narrow strip, north of Puttiala.

Puttiala (lat. 29° 20′—30° 50′; long. 74° 43′—76° 52′; area, 4,731 sq. m.; pop. 1,326,840).—This is far the most important of the group, the territory being among the most fertile in Sirhind, producing large quantities of grain, mostly exported to Lahore and Amritsir. The chief town is on the river Kosilla, which runs past the town in a very deep channel, but which is so swollen in times of inundation, that a great embankment has been found necessary to prevent the walls from being destroyed. It is a compact, well built town, cleaner than most of the Sikh cities. The condition of the state has been greatly improved under the present Raja. Within Puttiala are fragments of other much smaller states, of which Narba is one, and Kuslia another.

9. Independent Native Countries adjoining North-Western India.

Beyond India, to the west, is a large and important tract

of country, inhabited by races with whom it is absolutely necessary that the authorities of British India should occasionally come in contact. This country includes the two states of Afganistan and Beluchistan. Some notice of their physical features and population seems desirable in a work devoted to the geography of India.

Afganistan, or the land of the Afgans, is the northern part of the wide tract extending from India to Persia. It lies between lat. 28° 50′—36° 30′ N., and long. 62°—72° 30′ E. Its area about 225,000 square miles. From north to south it stretches down from the crest of the Hindu Cush, includes the whole drainage of the Cabul river and the table-land to the south, and is there terminated by the frontier of Beluchistan. It is bounded by the Suliman mountains on the east, and by Persia on the west. Much of it is mountainous and inaccessible, but it includes a succession of ridges and valleys, the valleys being irregular, and the ridges occasionally rising into lofty mountains, or expanding into plateaux. The highest summits are in the north, and exceed 20,000 feet, many of the passes across them exceeding 10,000 feet. The country is drained eastward to the Indus by the Cabul river, and westward into swamps and lakes. The former is the principal stream in the country, and falls into the Indus, near Attock, after a course of about 250 miles. Of the other streams, running southward, some are lost in swamps; some disappear altogether, being absorbed into the soil and never reaching the sea, except during the rainy season.

Four-fifths of Afganistan is a region of rocks and mountains, interspersed occasionally with well watered, fertile valleys, and in many places containing elevated table-land, yielding a scanty pasture. With a surface as rugged as Switzerland, it exceeds Spain in extent, and its climate brings to perfection many tropical productions in the lower parts; while the vegetation of the colder parts of the temperate zone prevails on the plateaux.

The valley of the Cabul is the most important part of the country. To the south is the fertile district of Logur, and to the north the Koh-i-Daman, also fertile and highly cultivated. To the east is the rich and beautiful vale of Jelalabad.

The mineral wealth of the country is considerable. It yields gold in many places. Silver, mercury, copper, antimony, and zinc are all present, some in abundance. The indications of copper are particularly dwelt upon. Iron is so abundant that it is unnecessary to indicate the localities. Coal has also been found.

The climate of Afganistan, though varying greatly in different parts, is on the whole characterized by dryness and great extremes of temperature. In Cabul the cold is intense, and snow lies on the ground for three months in winter. Even in a latitude lower than that of Spain or Italy, the severity of a Russian winter is endured. In Jelalabad, however, where the elevation is 2,000 feet above the sea, the winter is as mild as in Hindustan. The heat of summer is everywhere very great, and in some places higher than in Bengal. Even at Cabul, 6,000 feet above the sea, the thermometer ranges from 90° to 100°.

Much of this large country is irreclaimable desert, most of it is unenclosed, and nearly everywhere there are extensive wild tracts; but the number of wild animals is comparatively small, nor are the feline tribes ferocious. The goats and sheep are among the most valuable of the quadrupeds. There is a great variety of birds, and very few poisonous reptiles. Of forest trees, there are many of those common in Europe, and some others peculiar. The deodar (*Pinus deodarus*) flourishes on the mountain sides to a height of 10,000 feet. Species of oak, walnut, birch, and other trees grow at lower altitudes. The valleys yield all the ordinary Indian crops.

The Afgans are chiefly a pastoral race. In religion, they are Mahomedans, of the Sunni persuasion. They are tolerant to Christians, but quarrel with the Mahomedans of the Shia sect. The Beluchi language resembles the

Pushtu, which is of the same family as Sanskrit, though essentially distinct. In appearance the Afgans resemble the Jews, but their dress and manners are Persian.

The government in Afganistan is very peculiar and patriarchal; the people are bold and independent, and little inclined to submit to any control. They have no regular tribunals, and each tribe is practically free under its own Khan.

The chief towns of Afganistan are Cabul, Guzni, and Candahar. *Cabul* is situated on the river of that name, in a picturesque, well watered, and fertile district, producing the finest fruits. It is a fine and interesting city, the principal bazaar exhibiting much architectural ingenuity and great beauty. Much of the town was destroyed by the British on its capture in 1842. The houses are built of sun-dried bricks, with much wood, and the place is very subject to earthquakes. It is divided into districts, each of which is fortified. The site of the town is 6,396 feet above the sea, and it is subject to extreme cold in winter. Cabul is about 160 miles west of Peshawur and the English frontier. *Guzni* is an ancient and celebrated town and fortress, built on the western extremity of a range of hills rising from a plain. It is 7,726 feet above the sea. The fort is commanded by neighbouring hills. In consequence of its position, the cold is intense in winter. The population is small, but the bazaars large, and the town does a considerable amount of traffic. *Candahar* is the principal city of Western Afganistan. It is situated in a fertile and cultivated plain, well watered by canals from rivers that flow near it to the east and west. But at no great distance precipitous and rocky hills rise around it. It is enclosed by a mud wall. There are two principal streets, crossing at right angles in the middle. The buildings are poor, but the town is well supplied, though unhealthy. The population is very variously stated, but is believed to be 50,000. It is much mixed.

Afganistan communicates with Hindustan by the

celebrated Kyber Pass. It is very difficult, though only 3,373 feet above the sea, and has been obstinately contested. It commences near Peshawur, and extends about thirty miles to the plains of Jelalabad.

Beluchistan.—This country lies south of Afganistan, extending to the Indian Ocean, and bounded on the east by Sind, and on the west by Persia. It lies between 24° 50′ and 30° 20′ north latitude, and 57° 40′—69° 18′ east longitude. Its area is 160,000 square miles. Its coast line is regular and craggy, but not elevated. Towards the interior, however, there is rapid elevation. On the coast there are several roadsteads, but no good harbours. The interior is rugged, barren, and deficient of water. Its eastern side is crossed from north to south by the Hala mountain range, which approaches the Suliman range, but there are no lofty eminences. The whole country is described as a maze of mountains, except on the north-west, where it becomes part of the desert.

The rivers of Beluchistan are the *Bolan* and the *Mula* in the north, the *Hub* and *Purali* in the south, and the *Dusti* in the west; but they all dry up or are lost in the earth in dry weather, and in the wet season are destructive mountain torrents. None of them flow through regular and well defined channels. Along the whole 600 miles of coast which Beluchistan possesses, there is no stream which might not in dry weather be forded by a child.

The climate is extreme. The cold during winter is exceedingly intense, snow lying on the ground for two months in winter even in the fertile valleys, while in summer the heat is overpowering on the lower grounds. The country is said to be rich in mineral productions, copper especially being met with in large quantity. Lead and antimony have both been recorded. Sulphur has been extensively worked. There are mud volcanoes at Lus, near a place where iron ore is worked.

Part of Beluchistan, to the west, consists of a sandy desert, quite impassable in summer owing to the sand-storms, when the wind is so scorching as utterly to destroy animal life.

The inhabitants are chiefly Mahomedans of the Sunni persuasion, and are said to be very bigoted. The Beluchis appear to consist of a great admixture of races, of whom some prevail in one part and some in another. Some of them are active, hardy, and predatory, others devoted to agricultural pursuits. They are hospitable but indolent, and more fond of gambling and amusements than of any industrial pursuits.

The chief town of Beluchistan is *Kelat*, the capital, built on an eminence, and surrounded by a mud wall. It has some little trade, and is the seat of government. It is 6,000 feet above the sea, and its population is about 12,000. *Bela* is near the coast, about 100 miles north-west of Kurrachi. It contains 5,000 inhabitants, living in mud houses, and surrounded by a mud wall. Its streets are said to be neat and clean. *Kedge* is in the extreme west of the country, not far from the coast, on the Dusti. It has a fort, built on a high precipice, considered impregnable. The town was once large, but is now decayed.

Beluchistan is reached from Sind by the Bolan Pass, which is a succession of ravines and gorges, across the Hala chain. The crest of the pass is 5,793 feet above the sea; and there is little descent on the western side, as it merely reaches the top of the plateau. The ascent is difficult, the country very barren, and part of the road cut through high perpendicular hills. The total length is fifty-four miles, and the average ascent ninety feet in a mile. The lower part is very unhealthy in summer; and the pass is infested by robber tribes. It can be traversed by artillery.

CHAPTER IV.

CENTRAL INDIA.

Boundaries and Subdivisions:—1. *The Central Provinces.* General Account—Nagpur and Chuttees Gur Provinces—Nerbudda and Jubbulpur Provinces—Chota Nagpur Province. 2. *Hyderabad.* General Account—Assigned Districts of Hyderabad—The Nizam's Dominions. 3. *Malwa, or Central India.* Ali Mohun—Amjerra—Bopal—Bundelcund—Burwani—Dewas—Dar—Gwalior Territories—Indore Territories—Jabua—Jowra—Kurwai—Kothi—Myhir—Ocheyra—Omutwarra—Rewa—Rutlam—Shagur—Sohawal—Sitamow. 4. *Orissa States.* Jeypur and Hill Zemindars—South-west Frontier of Bengal States—Cuttack Mehals.

Boundaries and Subdivisions.—In the interior of the Peninsula, between the Lower Provinces of Bengal, the North-West Provinces, and Rajputana, on the north, and the Madras Presidency on the south, there is an extensive tract, nearly half of which is English, and the rest still governed by native princes, more or less subject to British superintendence. This country, a part of what is called "the Deccan" (*see* p. 157), lies between the Jumna and the Kistna rivers. It does not reach the sea, being separated from it by the Madras collectorates on the east, and those of Bombay on the west. Much of the territory is high table-land, breaking here and there into mountains; and, compared with other parts of the country, it is little cultivated. The total area is about 366,350 square miles, and the population is estimated at twenty-five millions. The name of Central India is fitly applied to this district, although, in the division into Presidencies, the British portion is included in that of Bengal. The following is the most convenient subdivision:—

		Area in square miles.
1. The Central Provinces (British)	. .	140,000
2. Hyderabad	Berar, or Assigned districts	17,334
	Nizam's dominions (Native)	78,000
3. Malwa, or Central India (Native)	. .	78,770
4. The Orissa States (Native)	. . .	52,234

At present there is access to most parts of the district by good roads, and the main railways of India reach and cross it. The part of the main line of the Great Indian Peninsula Railway from Bombay to Jubbulpur, now advancing to completion, will coast the whole of the northern extremity of the Central Provinces, and give direct communication both with Calcutta and Bombay. The important branch from Bosawul to Nagpur, actually open, crosses the Assigned district of Berar to the Central Provinces, and lays open an important cotton-growing district to the sea at Bombay. The other part of the main line of the same system, branching at Calliani and completed to Sholapur, is being continued through the south-westerly part of the Nizam's dominions to Raichur, where it will meet the north-western branch of the Madras railway. From Nagpur and Hyderabad there are good roads in several directions. There is a line of telegraph crossing the district from north to south, and it is proposed to construct a loop line through Berar.

The climate of the central part of the Indian peninsula is greatly affected by its geographical position. Situated on a table land, the country is not subject to the extreme and moist heat of Bengal or of the eastern coast. Very large districts, however, are covered by forest, and much of the country is inaccessible, and, perhaps in consequence of the forest, exceedingly unhealthy.

The natural resources of this part of India are very considerable. There is much coal and iron ore near Jubbulpur. Opium is largely cultivated in Malwa. The river valleys yield valuable crops of indigo, hemp, jute, and other fibres; while the forests contain trees yielding gums and other articles of export. Cotton is cultivated with advantage almost everywhere.

1. The Central Provinces.

General Account.—This group of provinces lies between 18° and 24° north latitude, and between 76° and 85° east longitude, reaching from Bundelcund in the north to the Madras Presidency in the south; and from the Bengal frontier on the east to independent Malwa and the Deccan in the west. The provinces extend about 550 miles from east to west, and 510 miles from north to south, occupying an area of 114,718 square miles. They are for the most part wild and rugged, with much forest and brushwood, these parts being thinly peopled; but they include considerable tracts well cultivated and thickly peopled. The whole country is crossed by the Satpura Hills.

The total population of these provinces is estimated at 9,104,511. The people are Mahrattas, Mahomedans, and Gonds, and there are many of the class called Banjaras, who are public carriers attending armies as camp-followers in time of war, and in peace transporting grain, cotton, salt, and other goods, from one part of the interior of India to another. The Gonds inhabit the hilly and forest lands, and are little advanced beyond the state of savages. They worship the goddess Kali, under the name of Mahadevi. The languages spoken are Mahratti in Nagpur, Hindi in Sumbulpur, and Telugu in the Godavery country. In the Gond district there are many dialects.

The seasons in this part of India are three—the hot, the rainy, and the cold. The hot season begins in April and lasts till the middle of June; the wet season then sets in and continues till October; and the cold season lasts from November to March inclusive. At all seasons the range of temperature is considerable. During hot weather the heat at noon averages 100°, while the mornings and evenings are comparatively cold. In the cold season the mean temperature is 40°; but the thermometer sometimes rises suddenly to 80°, and then as suddenly falls. Fogs and hailstorms occur at such times, and ice forms in winter.

The natural productions of the Central Provinces are cotton of the finest quality, rice, wheat, maize, millet, oilseeds, opium, sugar-cane, safflower, and indigo. Lac abounds in the forests. Fibrous, medicinal, and edible plants and trees are also found in great abundance; and there is much valuable timber, and many trees yielding resins, gums, and dyes. The mahowa-tree, which is spread over the Hill districts, yields food, oil, and an intoxicating drink. The mineral resources include iron ore, coal (chiefly on the Nerbudda), marble, and building-stone, gold, and diamonds. Oxen, buffaloes, sheep, and goats are the domestic animals. The forests swarm with wild animals of the kinds found elsewhere in tropical India.

There are few arts or manufactures, or other industries, beyond agriculture. Weaving, and the making of certain brass and copper utensils, and some rude iron manufactures, are the only exceptions. Agriculture is carried on in the rudest manner, especially in the forest districts, where ploughing is unknown. A piece of ground is selected on a slope covered with trees; the trees are cut down in winter, and the brushwood and grass burnt the following spring. In the beginning of June, before the rains, seeds are placed at the upper end of the slope; the rains carry down the seed and distribute it; and the result of this rude process is that a plentiful crop springs up.

The government of the Central Provinces is conducted by a Chief Commissioner appointed by the Viceroy of India, the method being that of the non-regulation system, as adopted in the Punjab. Under the Chief Commissioner are four Commissioners and seventeen Deputy Commissioners. The provinces are Nagpur, Jubbulpur, Chuttees Gur, and Nerbudda. The Central Provinces are included within the Bengal Presidency.

The condition of the people in the Central Provinces has, till lately, been exceedingly unsatisfactory. Owing to the entire absence of education, the want of means of communication, and long-continued bad government, the people have been among the most backward in India. Many of them were altogether barbarous, and sunk in the grossest superstition, even human sacrifices being offered by the Gonds to their terrible goddess. Crime of all kinds was common, and the lower classes dwelt in the forests in rude huts, and subsisted on the produce of the chase. Even the towns were dirty, ill-built, and unhealthy. Since the establishment of British authority matters have been improving. In 1818 the Saugor and Nerbudda country was ceded; in 1853 Nagpur was annexed, and soon afterwards Sumbulpur; and the whole was constituted a separate government in 1861. In 1862 an educational system was established, and already there are upwards of 1,500 schools, with some 50,000 scholars. The schools are as yet on a humble scale, but the improvement is marked. Means of communication have also greatly increased. Roads and bridges, canals, tramways, and even railroads, are already in existence; and the telegraph flashes communication from the capital to the very centres of darkest ignorance. The navigation of the Godavery has also been greatly improved. The Great Indian Peninsula Railway, open from Bombay to Nagpur, and from Calcutta to Jubbulpore, already taps the country at two most important points. The ordinary road from Nagpur to Calcutta is in excellent order.

These provinces are watered by tributaries of the Nerbudda and the Godavery. The Nerbudda is a river full of natural obstacles to navigation, with several large waterfalls, and runs through a country wild and unmanageable in every respect. During the rainy season its volume is enormously increased. It runs in a very direct course about 190 miles to Jubbulpur, and 225 miles further to Hoshungabad. Below this, for about eighty miles, the course is comparatively free from obstacles; but then occur rapids, and a fall of ten feet, which cause a stoppage to navigation. This state of the river continues to Juga, 346 miles above the mouth, and 455 miles from the source. After this it is navigable, but shallow.

The Godavery rises in the Deccan. It is only a group of feeders of the upper part of this stream that water the Central Provinces.

Nagpur and Chuttees Gur.—The southern group of provinces comprises Nagpur and the Chuttees Gur, each divided into several districts. It is watered by the Wein Gunga river, a tributary of the Godavery, and by tributaries of the Mahanuddy, and is separated from Hyderabad by the River Wurda. It is a tract of considerable elevation, the surface sloping from north-west to south-east, and the whole of the drainage being discharged into the Bay of Bengal, except two very small portions, which pass into the Arabian Sea. North of these provinces are the districts of Nerbudda and Jubbulpore, which reach the Vindya mountains, and rise to about 2,500 feet above the sea. Towards the south the surface is less rugged, and finally sinks into an extensive plain, about 1,000 feet above the sea. The division of Chuttees Gur, in the eastern part of the territory, rises in its northern part to 3,463 feet at Ummurkuntuk, where the Nerbudda takes its rise. Southward from this summit is the range of the Lanji Hills, about 2,300 feet above the sea. These hills divide the province into two basins, the northern draining into the Mahanuddy, and the southern into the Godavery. The south-eastern portion of the Nagpur province is a great wilderness, stretching away from the valley of the Wein Gunga down to the Godavery on the south, and almost to the range of the Eastern Ghats overhanging the littoral districts of Bengal and Madras. It is divided into two portions by the Idrawatty river, and the part to the north of that river is entirely uncultivated and unin habited. The whole tract is believed to abound with forests, but its climate is malarious and even deadly. The lower part to the south of the Idrawatty is also wild and unhealthy, and is described as an interminable and primeval forest, with a sprinkling of small villages. There are few roads or even paths through it, and it is inhabited chiefly by Gonds.

Although there are no natural lakes in the cultivated parts of the district, several very large reservoirs, artificially constructed by embankments, close up natural depressions among the hills, and take the place of lakes. One of these reservoirs is twenty-four miles in circuit.

The population of the district consists chiefly of Hindus, with a few Mahomedans, and nearly half a million of Gonds. The Mahrattas having been the ruling race, their language is used in the cities, and the Gond language is spoken by the people of that race.

Beyond the eastern extremity of the Nagpur province is a detached portion of territory, now a part of the Chuttees Gur. It formed the ancient raj, or native state, of *Sumbulpur*. It extends from lat. 21°—22° 5′; long. 83° 6′—84° 51′, embracing an area of 4,693 miles, and having a population of 274,000. The River Mahanuddy flows through it, and divides it into two unequal parts, the eastern part being mountainous and woody, the western level and depressed, with rich alluvial soil, producing abundant crops of rice, wheat, and sugar-cane. Gold and valuable diamonds have been found in the bed of the Mahanuddy.

Of the various towns in these two provinces, the following are the most important:—

Bustar, or *Jugdulpur*.—A town on the Idrawatty on the frontier of Jeypur. There is a fort adjacent on a peninsula formed by the river. Distance S.E. from Nagpur, 225 miles.

Chanda.—A town on the south-west frontier, near the territory of the Nizam, situated on the left bank of a small stream. It is a straggling place, fortified by a stone wall, from fifteen to twenty feet high, with round towers at intervals. The wall is six miles round. Within is a citadel. Distance S. from Nagpur, 85 miles.

Hingun Ghat.—A town on the River Wurna, a tributary to the Wurda. It is a place of considerable trade, and there is extensive cotton cultivation in the country round. From this neighbourhood are obtained specimens of silicified palm-trees, and other curious and interesting fossils. Distance S. from Nagpur, 45 miles.

Kampti.—A large British cantonment, nine miles from Nagpur. The climate is very hot, and it is subject to heavy hail-storms in the month of April. Hailstones have fallen measuring nine inches in circumference.

Nagpur.—The chief town of the Central Provinces. It is situated in a low swampy hollow, on the banks of the River Nag, and though there is some drainage into reservoirs, the place is muddy and wet during the rains. The town is about seven miles in circumference, but very straggling and irregular. There is only one good street, the others being mean, dirty, narrow, and almost impassable during the rains. There are so many trees interspersed among the houses, that at a distance the town looks like a forest. These trees interrupt vegetation, and render the place unhealthy. Most of the houses are built of mud, some being thatched and others tiled; but there are a few of large size, built of bricks and mortar, with flat terraced roofs. There are no good buildings whatever; even the palaces of the late Raja being a mere ugly pile of masonry, unfinished, and little ornamented, except by lofty pillars of carved wood supporting porticos. The manufactures of Nagpur include chintzes, coarse blankets, tent cloths, and copper and brass utensils. There is a considerable amount of trade carried on, and banking business is an important occupation. Owing to the recent opening of the railway to this town, and the increased means of communication, it has greatly improved in all respects.

Although Nagpur is situated in the very middle of the peninsula, about 350 miles from the Bay of Bengal, and

420 from the Arabian Sea, the mean annual rainfall is heavy, amounting apparently to sixty-five inches, of which five-sixths fall in the four SW. monsoon months (June to September inclusive). The range of the thermometer is small. The mean temperature is about 80°; somewhat higher than Calcutta and lower than Madras. The town is 930 feet above the sea. Distance ENE. from Bombay, 440 miles; W. from Calcutta, 605 miles; N. from Madras, 565 miles. The population is estimated at upwards of 100,000, of whom almost all are Braminists.

Close to the city, on the west, is the Ridge of Sitabuldi, running north and south. The northern is the higher part, but the whole length of the ridge commands the city in a military sense. It is probably owing to this range that the rains are so heavy.

Ruttunpur. — The capital of Chuttees Gur, in the eastern part of the district. It is a mere collection of huts, situated in a champaign country, abundantly watered by little rivers, full of villages, and having numerous reservoirs. It is 244 miles NE. of Nagpur.

Sumbulpur.—The capital of the eastern detached portion of the Chuttees Gur, bearing the same name as the town. It has no trade, though its situation on the river Mahanuddy would render it easy to communicate with the coast through Cuttack. It is very unhealthy.

Wyragur.—A small town on the left bank of the River Wein Gunga, a tributary of the Godavery, situated eighty miles SE. of Nagpur. Near it, in yellow earth, forming low hills, were formerly extensive diamond washings, but they have been discontinued. It is a place of some trade, but has a small population.

Nerbudda and Jubbulpur. — The northern division of the Central Provinces is an elevated table-land, 3,500 feet above the sea, extending into the mountain ranges of the Vindya and Mahadeo to the north, and of the Satpura range south of the Nerbudda. It declines towards the west to the valley of the Nerbudda, and towards the north it also includes some districts watered by the tributaries of the Godavery. Very large quantities of iron ore have been found at various places, and important beds of coal have been discovered. Iron is smelted at Punassa and other places. Good limestone and valuable sandstone, well adapted for building purposes, have been quarried. The population consists to a great extent of Gonds, perhaps the aboriginal race of this part of India. They lurk in the gloomiest recesses of the forests, and live on wild roots and fruits, wild honey, and game. They are even said to be cannibals, and they have certainly offered human sacrifices till within a very recent period. There is a considerable Mussulman population in some parts, and there are Bramins, Bundelas, Rajputs, and Mahrattas. The following are the principal towns of the Nerbudda and Jubbulpore provinces:—

Baitul.—A town pleasantly situated at the foot of the Satpura range, on a small river, a tributary of the Iowa river, which runs into the Nerbudda near Hoshungabad. Coal is obtained in the surrounding district. The town is old, and there is a fort. Distance S. from Agra, 370 miles; NE. from Bombay, 390 miles; W. from Calcutta, 677 miles.

Belhari.—A town formerly prosperous, but now ruinous. It is on a plain on the north-east frontier of the province, and there are some fine Hindu temples in the vicinity, as well as in the town itself. Distance NE. from Jubbulpore, 52 miles; SW. from Allahabad, 220 miles.

Chindwara.—A town on the route from Saugor to the city of Nagpur, situated in the mountainous tract called Deogur, above the Ghats, on an elevated tableland, having an open space of about four and a half miles in circumference on the summit free from jungle. Elevation above the sea, 2,100 feet. In consequence of its considerable elevation, its climate is one of the most agreeable and salubrious in India, and it is much visited for purposes of health and recreation.

Damouni.—A town on the frontier towards Bundelcund. It is surrounded by a loose wall. There is a fort on an eminence adjoining, strong and regularly built, with interior defences and magazines.

Damo.—A town near the frontier of Malwa, having a large bazaar, and well supplied with water. Distance from Calcutta, 775 miles.

Hoshungabad.—A town on the left bank of the Nerbudda, in the valley of the Nerbudda, in a district so remarkable for fertility, that it is commonly styled the garden of Central India. Besides rich crops, there are very valuable seams of coal in the neighbourhood. The town is irregularly built, and the houses much dispersed. It was founded in the early part of the fifteenth century, by Hoshung Shah, sovereign of Malwa. It has passed through several hands, and was finally ceded to the British Government in 1818. The Nerbudda at Hoshungabad is half a mile wide, and not fordable opposite the town. It is greatly infested with alligators. Near the town is a small cantonment. Distance W. from Calcutta, 924 miles.

Jubbulpur, also on the Upper Nerbudda. It is a large well built thriving place, favourably situated in a populous and highly cultivated country. The Nerbudda is there only 300 yards wide, and is fordable in the dry season. Around it are several reservoirs and small lakes, which are greatly swollen in the rainy season. The country near is particularly interesting to the geologist, presenting an unusual variety of porphyritic rocks, besides limestone, crowded with fossil bones of a tertiary period, including remains of elephants, and of several extinct quadrupeds. Excellent coal is found within the Jubbulpur district. Distance W. from Calcutta, 718 miles; NE. from Nagpur, 156 miles.

Mundla.—A town and fort on the right bank of the Nerbudda, once considered strong, but taken by the British in 1818, and now in ruins.

Nursingpur.—An important town on the branch of the Great Indian Peninsula Railway now in construction to connect Bombay and Calcutta. It is in the Nerbudda Valley, 50 miles below Jubbulpur.

Saugor.—A town built round three sides of a small but beautiful lake, about a mile long, which occupies the lower part of a basin surrounded with basaltic rock interspersed among sandstones. It is situated in a hilly district, and is a clean, well built, populous town, with many English residents; near the town are military cantonments, said to be unhealthy. There is a large fort here, now used as an ordnance depôt; and formerly there was a mint, but this is now transferred to Calcutta. An iron bridge has been constructed near the town, 200 feet in span, constructed of metal obtained in the neighbourhood, and made by native workmen. There is an English church and a resident chaplain. Distance NE. from Bombay, 500 miles; W. from Calcutta, 808.

Sohagpur.—This town, the principal place of a large district, is 170 miles E. of Saugor. It is little known.

Chota Nagpur (lat. 22° 25′—24° 45′; long. 82°—87° 10′; area, 25,284 sq. m.).—A portion of the ancient kingdom of Orissa, situated in the south-western corner of the Lower Provinces of Bengal, between the division of Burdwan and Central India. It is a hilly district, imperfectly known and thinly peopled. It consists for the most part of a table-land, between the valley of the Ganges and the Deccan, in some parts 3,000 feet above the sea. Numerous streams proceed from it in various directions, of which the Damuda* is the principal. The forests yield large quantities of valuable timber, dye-woods, and drugs; the mountains adjacent contain iron and copper ores. There are also stores of coal. The whole country is badly supplied with roads and other means of communication, and is far from the sea, so that although some parts are exceedingly fertile, the natural wealth cannot easily be rendered available.

Chota Nagpur is one of the three non-regulation pro-

* The Damuda is one of the tributaries of the Hoogly, and has a considerable course rising in the hills of Lohardugga. It run through a valley extremely rich in coal and iron ore.

vinces of Bengal; Assam and Cooch Behar, on the north-eastern frontier, being the other two. It is divided into four districts and governed by a Commissioner; there is also a Judicial Commissioner.

Lohardugga (area 10,314 sq. m.) is the largest of the districts of Chota Nagpur, and borders the province Behar. It is traversed by many torrents during the rainy season, and by one river, the Koel, which is permanent. It is a mountainous country, the mountains and valleys being covered with jungle, which abounds with valuable vegetable products, including the *sal*, and other trees, and remarkable for a species of wild cattle of gigantic proportions, the wild buffalo, the elk, the nilgau, and other interesting animals. There is coal and iron in the mountains. *Lohardugga* is the residential town of the province, and has some public buildings, but it is an inconsiderable place. An annual fair is held near it. The town of *Palamow* is situated near the Koel, amidst mountains containing coal and iron. It is likely to become important.

Maunboom, or *Pachete* (area 5,559 sq. m.), is an extensive district in the east of the province, adjoining Burdwan. A large part of it consists of a maze of mountains and ravines. It has a ruined town (*Pachete*) containing but few inhabitants.

Ramgur, or *Hazaribag* (area 7,021 sq. m.), is another large district in the northern part of the province. It occupies an extensive plateau of granitic and metamorphic rock in the northern part of the province. This plateau is 1,800 feet above the sea, thinly peopled, covered with forest abounding with wild animals, and traversed by many streams. The district is very unhealthy in summer. Lead and silver ores are believed to exist in the mountains. The towns are very unimportant at present. *Hazaribag* is small. It has a good bazaar, and was once a considerable town. *Ramgur* is still smaller.

Singboom (area 2,390 sq. m.) is a district in the S.E. of the province, adjoining the Cuttack Mehals. *Chaibassa* . . Near it copper is worked.

2. Hyderabad.

General Account.—South of the Central Provinces is the Deccan, a term which in its usual acceptation implies the tract of country in Southern India situated between the Nerbudda and Kistna rivers. Properly speaking, however, it includes the whole of the territory south of the Vindya mountains, comprehending the Valley of the Nerbudda and the narrow tract of low land forming a belt round the coast of the peninsula, besides the vast triangular expanse of table-land resting on each side on the Eastern and Western Ghats, and supported at its base by the Satpura, or sub-Vindyan range. On the west side the Ghats, or coast range, seldom exceed 3,000 feet, though towards the south they rise to 8,700 feet in the Neilgherries. The Eastern Ghats are much lower, not averaging more than 1,500 feet. The great table-land of Southern India, or the Deccan, thus understood, has a gradual slope eastwards, indicated by the course of the streams, which have their origin on the eastern slope of the Western Ghats, and make their way through fissures in the Eastern.

The Deccan, in its limited sense, occupies, however, a much smaller tract, a large part of which remains as an independent country, to a certain extent under British protection and supervision, under the name of "the Nizam's dominions." It is also called Hyderabad, from the name of the capital. This part of India occupies nearly the centre of the peninsula, between latitude 15° 10′ and 21° 45′ N., and longitude 74° 40′ and 81° 32′ E. It lies between the Wurda and Godavery rivers on the east, the River Kistna on the south, and the Bombay territory on the west. It comprises 95,334 square miles of country, and its population is nearly twelve and a quarter millions. It is on the whole a table-land; but is not without undulations, rising at Beder to about 2,000 feet above the sea. A little beyond that district the table-land reaches 2,700 feet.

This country is to a large extent covered with brushwood, and uncultivated, but there is no extent of forest trees in any of the jungles. Where irrigated and cultivated, the soil produces cotton, wheat, and oil-seeds in great abundance, and extensive plots of date-palm and palmyra-trees are found everywhere. European vegetables are raised in perfection at all the military stations, and fruits, as strawberries, figs, grapes, and peaches, ripen perfectly. The climate is, on the whole, pleasant and healthy. During a great part of the year the temperature is moderate, and is described as a delightful medium between the extremes of heat and cold experienced in the northern parts of India. In the cold season, from the middle of November to the middle of February, the thermometer stands at 74°; in the hot months, from the latter period to the end of May, at 91°; and in the rains, from the early part of June till October, at 80°. The average rainfall is very small, not exceeding thirty-two inches, and occurs at the change of the monsoons, the south-west monsoon (beginning of June to beginning of October) bringing the heaviest rains. At mid-winter, in the northern part of the country, the mornings are very cold, and ice is formed, but the days are hot. Towards the close of the monsoons, fevers and agues are common, but not fatal, except near marshy jungles. The water obtained from wells is generally bad, and productive of disease.

The coal in the Nerbudda Valley occupies an area 140 miles in length, with a varying breadth of from 20 to 80 miles, and licences have been granted to work it. Coal has also been found near Chanda, on the Nizam's side of the Wurda, reported to be of good quality. It is probable that along the line of the Wurda, for a distance of 90 miles, the strata are coal-bearing.

Hyderabad is crossed by several rivers. The principal are the Godavery, with its tributaries, the Dudna, Manjara and Pranhita; the Wurda with its tributaries; and the Kistna, with its feeders, the Beema and Tungabudra.

Hyderabad is governed by a potentate called the Nizam, or "Regulator," the title of the governor of the

Deccan at the time when this, as well as many other rulers of provinces, constituted themselves independent princes, on the dissolution of the Mogul Empire after the death of Aurungzebe, about the middle of the last century. At this time the French were still powerful in India, and, after the decease of the first Nizam, some of the claimants to the succession appealed to the French, others to the English. The latter proved the most efficacious allies, and, after the wars with Hyder and Tippu, the Nizam Ali established his position. But a large debt to the English, for military assistance, had by this time accumulated, and as there appeared no probability of this debt being repaid in any other way, certain portions of the territory were assigned for payment. These are called "the assigned districts." The rest remains under the political control of the descendants of Nizam Ali.

Assigned Districts of Hyderabad (lat. 19° 30′—21° 45′; long. 76°—79° 10′; area , 17,334 sq. m.; pop. 2,231,565).—This portion of the northern extremity of Hyderabad, held in pledge for debts incurred, is now under regular British government. It consists chiefly of the valuable provinces of Berar. On the south it is bounded by the Payne Gunga river, and on the east is separated from Nagpur by the Wurda branch of that river. The Tapti, on the north, separates it from the Central Provinces. It is watered by numerous branches of these streams, and is crossed by the Nagpur branch of the Great Indian Peninsula Railway. It occupies an important position between the British district of Candeish, in the Bombay Presidency, and the province of Nagpur, in Central India. As a cotton-growing district it is one of the most important in India. Although the management is now entirely British, the sovereignty still remains with the Nizam. The country is at present divided into two Commissionerships (East and West Berar), with two districts each.

Umrawutti is the principal town, and a station on the line of railway. It has long been a place of great commercial interest in reference to the cotton trade of Central India, and has been in direct communication with Bombay for this purpose; but the railway now opened has added greatly to its importance. It is a large and growing place, 250 miles NE. of Bombay, and 245 miles N. of Hyderabad. (Pop. 23,410.) *Akowla*, or *Akola*, is a considerable place, close to the railway. It has high walls, well built, and its ruins indicate that it was formerly a place of some extent and importance. It is now likely to improve. It is about fifty miles nearer Bombay than Umrawutti, and its population in 1867 exceeded 14,000. *Akote* is another town of the same size. *Ellichpur* was the former capital, and is the largest town. Its population is 28,000. It is surrounded by high walls. *Maiker*, *Gotmal*, and *Meil* are other towns. At the time when this valuable country was brought under British administration in 1853, it was in a miserable state, the condition of the cultivators being most wretched, and violent crimes common. There were few roads, no bridges, and no police. Ruined irrigation works have been now restored, reservoirs made, railroads and other roads opened, bridges constructed, hospitals, schools, and police established, and many improvements effected.

The Nizam's Dominions (lat. 15° 10′—20° 40′; long. 74° 40′—81° 32′; area, 78,000 sq. m.; pop. 10,000,000). The general features of the country have been already described, and the following description is limited to the principal cities and towns.

HYDERABAD, the capital of the whole country and of the Deccan, is a fortified city on the River Musey, a tributary to the Kistna, in a wild but picturesque district, with granite hills and isolated blocks of granite jutting out of the ground through and amongst the vegetation. It is about four miles in length and three in breadth. The streets are narrow, crooked, ill paved, and dirty, and the houses poor, and chiefly built of wood. But the palace and numerous mosques, rising above the surrounding buildings, give it an air of grandeur, which is much strengthened by the pile of buildings erected in the British Residency, and the number of gardens and trees within the walls give the whole the character of a city built in a vast park. In the environs, also, are many fine gardens, containing gorgeous pavilions belonging to the nobles. There is a handsome stone bridge across the river. The population comprises a restless, reckless mass of Rohillas, Arabs, Afgans, Patans, and others, estimated at 200,000. Elevation above the sea, 1,800 feet. Distance NW. from Madras, 389 miles; SE. from Bombay, 449 miles; SW. from Calcutta, 962 miles.

Assaye.—An insignificant village, at the confluence of the rivers Kistna and Juah, in the tongue of land or doab between them. It is celebrated for the great victory gained here on the 23rd September, 1803, by Major-General Wellesley, afterwards Duke of Wellington, with 4,500 men, over 50,000 Mahrattas, of whom 10,500 were disciplined and commanded by European officers.

Aurungabad.—A decayed walled city on the Dudna, a tributary of the Godavery. Pop. 60,000. The view from the east is pleasing, trees being interspersed among the houses, and a tall mausoleum rearing its dome and minaret above the other buildings. It contains the palace of Aurungzebe, and a mausoleum built by him. The tomb is a bad resemblance of the Taj Mehal. Distance NW. from Hyderabad, 270 miles.

Beder.—A large town near the Manjara, a tributary of the Godavery, situated on a table-land 2,360 feet above the sea, and 100 feet above the surrounding country. It was formerly the capital of a principality of the same name. It is surrounded by lofty walls. It is now chiefly noted for various manufactures of a kind of bronze (twenty-four parts tin and one of copper), coloured black, and ornamented with designs in gold and silver. Distance NW. from Hyderabad, 75 miles.

Bolarum.—A military cantonment for the troops of the Nizam, 11 miles N. of Hyderabad, and 1,890 feet above the sea, on an elevated plateau of granite. It is very healthy.

Daverkonda.—A town on a hill, near one of the tributaries of the Godavery, about fifty miles south of Hyderabad. It is well supplied with water.

Dowlatabad.—A town and rock-fortress, ten miles NW. of Aurungabad, near the north-west frontier of the Nizam's dominions. The original name was *Deogur.* The town is at the base of the rock, an isolated cone of granite rising 500 feet above the plain. For nearly a third of the height the rock is scarped, and presents all round a perpendicular cliff. Access to the summit is obtained by a narrow opening excavated in the rock, leading to a large vault, whence a narrow winding passage, also cut out of the rock, leads to the top. At the base of the hill is a deep ditch, crossed only by one narrow causeway.

Eidgir.—A town on the left bank of the Beema, a tributary of the Kistna, 100 miles SW. from Hyderabad.

Ellora.—A decayed, town thirteen miles NW. of Aurungabad, formerly a place of some note. It is supposed to have been built in the tenth century. In the neighbouring mountain are the excavated temples of Ellora, consisting of a group of caves at irregular distances from each other, and at different levels, opening out of the face of a cliff looking towards the west, and rising to considerable elevations at the two extremities. The sculptures in these caves are among the most remarkable works ever produced. They all refer to early Hindu mythology.

Golconda.—A fort and ruined city (the capital of an ancient kingdom) situated seven miles W. of Hyderabad, and now used as a state prison. About 600 yards from the fort are the splendid tombs of the kings of Golconda, who ruled over this territory in the sixteenth and seventeenth centuries. They form a vast group in an arid, desert, rocky ground, the stern features of which heighten the impressiveness and grandeur of these astonishing buildings. They are chiefly constructed of grey granite. The diamonds formerly obtained at Punna, were cut and polished at Golconda; but no diamond mines exist near the town.

Jaulna.—A British cantonment in a dreary barren country, about forty miles east of Aurungabad. The climate is very well adapted to food-crops of all kinds. *Jaulna* is an old town on the Kundulka, two miles SW. of the cantonment. It was formerly large and flourishing, and was a place of manufacture for silks. It still has 10,000 inhabitants. The houses are well built. Opposite is the town of *Kaderabad*, with 7,000 inhabitants.

Kulburga.—A large town and military station, 110 miles from Hyderabad.

Kummummut.—A town on the left bank of the Munyair, a considerable tributary of the Kistna. It is the chief town of the district of *Palun Shah*, held by an officer who is almost independent of the Nizam. Distance E. from Hyderabad, 110 miles.

Mulkair.—A town on an important tributary of the Beema. Distance W. from Hyderabad, 86 miles.

Nandair.—A town on the left bank of the Godavery. It is a place of pilgrimage for the Sikhs, who pay their devotions at the tomb of a saint assassinated here. There is a Sikh college. Distance N. from Hyderabad, 145 miles.

Nelgunda.—A town at the foot of a granite hill, surmounted by a fortress 1,000 feet above the plain.

Nirmul.—A large town, nine miles north of the Godavery, on the road from Hyderabad to Nagpur. Distance N. from Hyderabad, 120 miles.

Secunderabad.—One of the chief British cantonments in India. It is six miles north of Hyderabad; and the barracks and quarters extend upwards of four miles in a straight line from east to west. There is a large reservoir in the neighbourhood, three miles in length by two in breadth, called the Husain Sagur tank. The surrounding country is granite. Elevation above the sea, 1,837 feet; distance NW. from Madras, 398 miles.

Wurungal, or *Warungul*.—The ancient capital of Telingana, but now a decayed place. The four gateways of the Temple of Siva still continue in a state of tolerable preservation. Distance NE. from Hyderabad, 86 miles.

3. Malwa, or Central India.

The portion of the Indian peninsula called Malwa is a table-land of uneven surface, elevated from 1,500 to 2,000 feet above the sea-level, bounded on the west by the Aravalli range of mountains, on the south by the Vindya chain and the Central Provinces, on the east by Bundelcund, and on the north-east by the Valley of the Ganges. It occupies a position between the Central Provinces and the North-Western Provinces of Bengal, both already described. It is divided into a number of principalities held by native chiefs, all of them being under the political superintendence of an Agent appointed by the Viceroy, who resides at Indore.

The whole extent of the various states is nearly 80,000 square miles. For the most part, they are well watered and exceedingly fertile, enjoying a fine climate; and they contain a large population.

Malwa was formerly a powerful kingdom, preserving its independence through a line of kings for 130 years. It was then annexed to the imperial dominions by Akbar, and continued so till its subjugation by the Mahrattas. It afterwards fell under the power of the Pindaris, an infamous tribe of plunderers, who were forced by the Marquis of Hastings to resort to more honest means of obtaining a livelihood than they had been accustomed to. They are kept in order by the Bheels, a corps embodied in 1840, and supported by the various princes.

The following are the principal states. There are some others, but they are very small and unimportant:—

Ali Mohun, or **Rajpur Ali** (lat. 22° 2′—22° 30′; long. 74° 16′—74° 44′; area, 708 sq. m.; pop. 69,384).—A small state between the Vindya Hills and the Nerbudda, formerly belonging to the Raja of Dar. There is a large and well built town, named *Rajpur*, with a good bazaar.

Amjerra (lat. 22° 16′—22° 47′; long. 74° 40′—75° 15′; area, 584 sq. m.; pop. 57,232).—A little state in the south-western corner of Malwa. It yields much opium, besides Indian corn, millet, cotton, and sugar. It has a small town, well supplied, situated in a valley open to the north, at an elevation nearly 2,000 feet above the sea. It is twelve miles west of Dar.

Bopal (lat. 22° 32′—23° 46′; long. 76° 25′—78° 50′; area, 6,764 sq. m.; pop. 663,656).—This state is bounded by the River Nerbudda to the south. The Vindya Hills, which enclose the valley of the Nerbudda, bound it to the north, and terminate in a plateau sloping gently to the north. This plateau is crossed by the Betwa, which is one of the tributaries to the Jumna, and by several small feeders of the Betwa. Bopal is governed by a Nawab, who is now a Begum, and much of it is held by chiefs. The chief town is *Bopal*, two miles in circuit, and surrounded by a wall of masonry, much dilapidated. Outside the town is a large commercial quarter, with wide straight streets. The residence of the Nawab is at *Futtegur*, on the south-west of the town. Beyond this is a lake four and a half miles long, deep, but apparently artificial, and full of fish and alligators. *Sehore* is a considerable town.

Bundelcund, or the Bundela country (lat. 23° 52′—26° 26′; long. 77° 53′—81° 39′; area of native states, 8,394 sq. m.; pop. 1,142,000).—This extensive district is partly British, and partly governed by native princes. The British portion includes Banda (p. 90), Humirpur (p. 91), Jaloun, Jansi, and Lullutpur (p. 101). The country consists of plains, diversified by mountains, which have been classed into three ranges—the Bindachal, the Punna, and the Bandair. From these, numerous streams flow into the Jumna. There are considerable mineral resources, among which are diamonds in Punna, and coal elsewhere. Iron is found everywhere. The climate is not unhealthy to the natives, but very fatal to Europeans in some places.

The following are the native states:—

Ajygur (area, 340 sq. m.; pop. 45,000). It contains a hill fort on an isolated granite summit. Within it are ruins of temples containing very remarkable sculptures.

Allypura (area, 85 sq. m.; pop. 9,000).

Baoni (area, 127 sq. m.; pop. 18,000).

Behut (area, 15 sq. m.; pop. 2,500).

Beri (area, 30 sq. m.; pop. 2,500).

Berounda (area, 275 sq. m.; pop. 24,000).

Bijawur (area, 920 sq. m.; pop. 90,000).

Bijna (area, 27 sq. m.; pop. 2,800).

Bysonda (area, 8 sq. m.; pop. 8,000).

Churkari (area, 880 sq. m.; pop. 81,000).

Chutterpur (area, 1,240 sq. m.; pop. 120,000).—The chief town is a thriving place, with manufactures, picturesquely situated.

Durwae (area, 18 sq. m.; pop. 8,000).

Duttea (area, 850 sq. m.; pop. 120,000).—The town is interesting, with architectural relics.

Goriar (area, 76 sq. m.; pop. 24,000).

Gorowli (area, 50 sq. m.; pop. 5,000).

Jigni (area, 27 sq. m.; pop. 2,800).

Jusso (area, 180 sq. m.; pop. 24,000).

Kampta (area, 1 sq. m.; pop. 300).

Logasi (area, 29 sq. m.; pop. 3,500).

Muckri (area, 10 sq. m.; pop. 1,600).

Nyagaon (area, 30 sq. m.; pop. 5,000).

Nygowan (area, 16 sq. m.; pop. 1,800).

Oorcha, or *Tehri* (lat. 24° 25′—25° 35′; long. 78° 23′—79° 22′; area, 2,160 sq. m.; pop. 240,000).—The largest of the native states of Bundelcund. The northern extremity is crossed by the Betwa river, a tributary of the Jumna, and it is surrounded by small states. The Raja is the head of the Bundela race. The chief town is three miles in circuit, with a wall of unhewn stones. It is on a rocky eminence, and contains a temple ornamented with lofty spires.

Pahari (area, 4 sq. m.; pop. 800).

Pahra (area, 10 sq. m.; pop. 1,600).

Paldeo (area, 28 sq. m.; pop. 3,500).

Punna (lat. 23° 52′—25° 5′; long. 79° 50′—80° 45′; area, 688 sq. m.; pop. 67,500).—A small state, south of Oorcha, celebrated as having contained in former times very valuable and extensive deposits of diamond matrix, which extend from twelve to twenty miles north-east of the town of Punna. There are some mines close to the town. The ground at the surface and a few feet below consists of ferruginous gravel, mixed with reddish clay, containing a few diamonds. From twelve to forty feet below this is the real diamond matrix—a conglomerate of pebbles, of quartz, jasper, hornstone, Lydian stone, and other minerals. These are pounded and washed. Below the matrix is a sandstone 400 feet thick; and there are said to be indications of coal underlying the whole mass. The town of Punna is in ruins. It was formerly well built and well paved, and is crowded with Hindu temples. The ruined palace of the Raja is very beautiful. The whole place is inhabited by monkeys.

Purwa (area, 12 sq. m.; pop. 1,800).

Sumtur (area, 175 sq. m.; pop. 28,000).

Surela (area, 35 sq. m.; pop. 4,500).

Tehri (see *Oorcha*).

Tehri Futtepur (area, 36 sq. m.; pop. 6,000).

Turaon (area, 12 sq. m.; pop. 2,000).

Urcha (see *Oorcha*).

Burwani (lat. 21° 41′—22° 9′; long. 74° 29′—75° 22′; area, 1,380 sq. m.; pop. 22,217).—A hilly district, the patrimony of a Bheel chief. It extends along the left bank of the Nerbudda, and is within the Satpura range of hills. The country contains much fine timber, and is well watered; but is only partially cultivated, and the population is scanty. The town of *Burwani* is two miles from the Nerbudda. It is surrounded by a double wall.

Dewas is made up of scattered portions of territory of trifling dimensions. There are altogether about 256 square miles, and a population of 25,000. It is ruled conjointly by two Rajput chiefs, whose ancestry is very distinguished.

Dar.—Another larger state, whose Raja's territory includes many detached patches at some distance from each other. The whole area is 2,091 square miles, and the population about 125,000. Much of the land is very fertile, producing abundant crops of rice, wheat, millet, maize, pulse, oil-plants, sugar-cane, tobacco, opium, ginger, cotton, hemp, turmeric, and excellent vegetables. The Raja is of very ancient family. There is a chief town, also called *Dar*, 33 miles W. of Mow, and 183 E. of Baroda. It is nearly three miles and a quarter long, and half a mile wide. It is surrounded by a mud wall, and has many striking buildings, among which are two large mosques of red stone. Water is abundant. Outside the city is a fort on an eminence forty feet above the plain. The city, once very fully peopled, is now almost deserted.

Gwalior Territories.—These territories, the possessions of the Sindia family (hence sometimes called SINDIA), have a singularly irregular outline, and consist of several detached districts, well watered by the Chumbul, the Kali Sind, the Tapti, and the Nerbudda, and numerous tributaries of these streams. The extreme points of the territory are included within lat. 21° 8′—26° 50′; long. 74° 45′—79° 21′; and the area of the whole comprises about 33,000 square miles, with a population of 2,500,000. The northern part of the country, which is of moderate elevation, and either rocky or sandy, has a very unhealthy climate in the wet season, when the air is always at the point of saturation; but during the dry and hot seasons this is not the case. The middle, southern, and western parts have a mild and equable climate, owing to the elevation of the surface. The nights are always cool and refreshing. The people consist to a great extent of Bramins, of whom there are very numerous tribes. There are also Mahrattas, Jats, Rajputs, Bats, and Charuns. The rulers are Mahrattas.

The founder of the ruling family of the Gwalior states was Ranoji Sindia, belonging to the tribe of cultivators, and a domestic of the Peishwa, or ruler, in the beginning of the last century. He rose from that to be a Mahratta chief of considerable importance, and was succeeded in 1750 by one of his sons, who attained still greater power. At the death of this son, in 1794, the territories of the Sindia family included Candeish, in the Bombay Presidency, Agra, Delhi, and the finest parts of the Doab. His successor was defeated at Assaye and lost a large part of his possessions, and died childless. After an interval of disturbed government, a British force marched into Gwalior, in 1843, to restore order. In 1853 the present Maharaja was entrusted with the administration, and the country has since been well governed.

Bilsa, on the Betwa, is celebrated for some curious monuments of antiquity in its neighbourhood. These consist of hemispherical constructions of vast size, connected with the worship of Buddha. The tobacco grown near Bilsa is regarded as the finest in India. Distance E. from Oojein, 134 miles; S. from Gwalior 190 miles.

Gwalior, the capital, is a large old town, of 50,000 inhabitants, irregularly built, and very dirty, but with one very beautiful building—a mausoleum. The town is built along the eastern base of a completely isolated rock of sandstone capped with basalt, about a mile and a half long, and 300 yards wide, the site of one of the celebrated hill-forts of India. Around are several hills, at a distance of from one to four miles. The height of the basaltic capping is 342 feet above the plain, at the highest point. The fort is entered by a steep road, succeeded by steps cut in the rock, but of so moderate an acclivity that elephants easily make their way up. There is a succession of seven gates before the summit is reached, and all are properly defended. It is said that 15,000 men would be required for the defence. The fort is very ancient, having been built in the eighth century. It has been frequently besieged and sometimes taken, more than once by stratagem, at other times by treachery. In 1779 it was scaled and taken by a sudden surprise by the English, and again in 1858 it was taken by Sir H. Rose. Distance S. from Agra, 65 miles; NW. from Calcutta, 772.

Mundesor, on a tributary of the Chumbul, eighty miles NW. from Oojein, is a town with a well supplied bazaar.

Neemuch.—A town and British cantonment near the Mewar frontier. It has a good bazaar. The camp stretches two and a half miles, with a breadth of a mile, and there is every convenience. The climate is exceptionally favourable and pleasant. There are no manufactures.

Oojein.—A city on the right bank of the Sipra. It is oblong, and surrounded by a stone wall six miles in circuit. The houses are much crowded, and are built with wooden frameworks filled with brick. There are four mosques and many Hindu temples. Outside the town are many beautiful gardens. About a mile to the north of the town are the ruins of the ancient capital of Malwa, destroyed by some unknown cause. This town was one of the seven sacred cities of the Hindus, and the first meridian of their geographers. It was called *Avanti*, or *Visala*. Distance from Gwalior, 260 miles.

Indore Territory.—This territory also consists of a number of widely separated tracts of country, covering altogether an area of 8,318 square miles, with a population of 576,000. The northern parts are watered by the Chumbul and its tributaries. The southern form part of the valley of the Nerbudda, and are crossed by the Vindya and Satpura mountain chains. They are the possessions of the Holkar family; the founder of the family was a cultivator of Hul, a village of the Deccan. He was born in 1693, and before he attained the age of sixty he had acquired the general management of Mahratta affairs in Malwa, and possessed the district of Indore. He died in 1767, and was the most distinguished of the military commanders of the Mahratta race. For thirty years the country was well governed by the widow of his only son, and another of the tribe who was military chief. The authority was assumed by one of the sons of this chief, who at first distinguished himself against the English. After much fighting, hostilities were terminated by a decisive battle, gained by the English, in 1817, at Mahidpur. The territory is at present governed by a Maharaja, the English engaging to maintain a force for the preservation of internal tranquillity, and for defence against foreign aggression.

The population of Holkar's dominions is estimated at about a million. Mahrattas are the ruling tribe; but there are many other classes of Hindus, a few Mahomedans, and many Gonds and Bheels. Indore is especially the country of the Bheels, who were once the most wild and savage tribes in India, living in forests on wild vegetables and game, and plundering their neighbours. They have been converted into excellent soldiers by the English.

The country is fertile, producing wheat and other grains, opium, sugar-cane, tobacco, and cotton, in abundance. Opium is a very important crop. The following are the principal towns:—

Agur.—A large town in an open plain near a reservoir. Pop. 30,000. It is 1,600 feet above the sea.

Banpura.—A town on the River Rewa, sixty miles E. of Neemuch, and the same distance S. of Kota. It is at the foot of a range of hills, and is surrounded by a wall. Population 20,000. Elevation above the sea 1,344 feet. There is a fine but unfinished palace within the walls.

Indore.—The capital and chief town of Malwa, and the

station of the British Resident. It is small, and irregularly built of sun-dried bricks, and contains a few mosques and temples, of no interest. The palace and the house of the Resident are surrounded with groves and gardens, and form a pleasing scene. The city is square, each side about 1,000 yards. Distance W. from Calcutta, 1,030 miles; NE. from Bombay, 377; SW. from Agra, 402.

Mow.—A town and British cantonment, thirteen miles W. of the town of Indore, 2,019 feet above the sea, and situated on an eminence near the Gumbir river. The cantonment is large, convenient, and healthy, and beautifully situated.

Mundlaisir.—A town on the right bank of the Nerbudda, here 500 yards wide. It is surrounded by a mud wall, and has a bazaar. It has a small fort. This town belongs to the British Government. Distance NE. from Bombay, 334 miles; NW. from Nagpur, 327 miles; SW. from Indore, 14 miles.

Rampura.—A town on the north bank of the river Taloyi, formerly the capital and residence of the Court. It is of considerable size, is surrounded by a wall, and has a good bazaar. North-east of the town is a Hindu temple, a place of pilgrimage in the month of April. Distance N. from Indore, 100 miles; from Oojein, 95 miles.

Jabua (lat. 22° 50′—23° 11′; long. 74° 18′—75° 2′.—A state, inhabited by Bheel tribes, regarded as part of Indore, adjoining Amjerra and Banswarra. *Borai* and *Jucknowda* are included. The town of *Jabua* is beautifully situated in a valley at the eastern base of a ridge of hills. It is enclosed by a mud wall. There is a fine lake south of the town, on the banks of which is the Raja's palace, which is fortified.

Jowra (lat. 23° 32′—24° 10′; long. 74° 53′—75° 35′; area, 872 sq. m.; pop. 85,456).—A small territory belonging to a Patan chief entitled Nawab of Jowra. It has a small town of the same name on the river Piria, with abundant supplies. The river is crossed by a handsome stone bridge.

Koti (100 sq. m.; pop. 30,000).

Kurwai (lat. 23° 50′—24° 12′; long. 78° 2′—78° 15′; area, 200 sq. m.; pop. 19,600).—A small state with a considerable town on the Betwa, and a strong fort. The inhabitants are Patans, and its chief formerly held a large territory. Opposite the town, on the other side of the river, is *Boraso*, also a considerable place.

Myhir (1,026 sq. m.; pop. 100,000).—The town is large, and has a good market.

Ocheyra (436 sq. m.; pop. 120,000).

Omutwarra (lat. 23° 28′—24° 9′; long. 76° 19′—77° 11′; area, 1,348 sq. m.; pop. 132,104).—A small state in the heart of Malwa, subdivided into *Nursingur* and *Rajgur*, and governed by two chiefs. There are small towns, one in each division.

Rewa, or **Bagelcund,** and **Mukundpur** (lat. 23° 20′—25° 10′; long. 80° 40′—82° 52′; area, 9,827 sq. m.; pop. 1,200,000).—A mountainous country between the North-Western and Central Provinces of India, rising in three successive plateaux from the valley of the Ganges. The first range is only about 500 feet above the sea, the second from 900 to 1,200 feet. The third, called the Kaimur range, is of much greater altitude. The lower range is sandstone, and barren; the second well cultivated, and produces grain. The third is exceedingly fertile. The produce is wheat, barley, pulse, and cotton. Coal and iron are found in large quantity. There are several rivers, none of them navigable. The country is well supplied with roads; and there is now a main line of railway at no great distance. The population—the Bagel race—consist, for the most part, of Rajputs. The town of *Rewa* is built on the bank of a tributary of the Tons;

it is small, and has an appearance of poverty, but is surrounded by three lines of rampart.

Rutlam (lat. 23° 2′—23° 36′; long. 74° 42′—75° 18′; area, 936 sq. m.; pop. 91,728).—A state situated in the western extremity of Malwa, bordering on Rajputana. There is a large well built town of the same name, with good bazaars, containing 10,000 inhabitants. It is governed by an influential, though poor, Rajah.

Shagur (676 sq. m.; pop. 30,000).

Sohawal (179 sq. m.; pop. 80,000).

Seeta Mow (area, 208; pop. 20,384).

4. Orissa States.

A number of states, of which Jeypur is the largest, between the Central Provinces (p. 146), and that part of the Madras Presidency called the Northern Circars. It is part of the ancient kingdom of Orissa (*see* note, p. 49).

Jeypur (lat. 17° 15′—19° 45′; long. 81° 28′—83° 56′; area, 13,041; pop. 400,000).—A tract of country in the ancient territory of Orissa, including a number of small states called the "Hill Zemindars." It is a wild rugged district, destitute of roads and other evidences of civilization, and inhabited by the tribes called Konds. The climate is very unhealthy. The Konds dwell in villages in the forests, each village consisting of two streets, composed of a double row of timber-built huts with thatched roofs. Many of the villages are stockaded. The people are warlike, wear very little clothing, and their chief occupation is hunting. Their agriculture is of the rudest kind. Schools have lately been introduced among them, and it is understood that their condition is improving, and that they are becoming more civilized. The British Government is represented in Jeypur by a Political Agent appointed by the Viceroy.

South-West Frontier of Bengal States.

Bombra (area, 1,244 sq. m.; pop. 55,980).—It adjoins Sumbulpur to the east, and is crossed by the Brahminy river, which connects with the delta of the Mahanuddy.

Bonie (area, 1,057 sq. m.; pop. 47,565).—A state situated to the north of Bombra, inhabited by very uncivilized races.

Bora Samba (area, 622 sq. m.; pop. 47,990).—A small state near Sumbulpur, situated on an elevated table-land, inhabited by savages. The climate is very cool.

Burgur (area, 399 sq. m.; pop. 17,955).—A very small state adjoining Ryegur.

Gangpur (lat. 21° 50′—22° 37′; long. 83° 31′—84° 57′; area, 2,493 sq. m.; pop. 112,185).—A vast jungle, affording admirable sport, but of little value to the human population. The soil is said to be rich.

Jushpur (area, 617 sq. m.; pop. 27,765).—A high table-land, overrun with jungle, but yielding rice, grain, and oil, in the cultivated parts.

Kerial, or **Bokur** (area, 1,512 sq. m.; pop. 68,040).—A wild uncultivated country, with a savage people. The town is *Kerial*.

Korea (area, 2,225 sq. m.; pop. about 100,000).—A wild state, in the north-western part of Orissa, near Rewa, inhabited by savages.

Nowagur (area, 1,512 sq. m.; pop. 68,040).—A badly governed state, near Berar.

Patna (area, 1,158 sq. m.; pop. 52,110).—A badly governed state, near Sumbulpur, crossed by two streams.

Phulgur (area, 890 sq. m.; pop. 40,050).—A level district, at a considerable elevation, with a good but neglected soil. The country is overrun by wild buffaloes.

Ryegur (area, 1,421 sq. m.; pop. 63,945).—A plain country, high and wild, but improving. The town is prettily situated.

Sarungur (area, 799 sq. m.; pop. 35,955).—A small state, separating Sumbulpur from the main portion of the Chuttees Gur.

Sirguja (lat. 22° 34′—23° 54′; long. 82° 40′—84° 6′; area, 5,441 sq. m.; pop. 316,252).—A large state adjoining Chota Nagpur. The country is rugged and mountainous, rising 500 to 600 feet above the adjacent table-land. It is crossed by two streams flowing towards the north; they are torrents in the rainy season, but generally shallow. The forests abound with the wild animals of the Indian jungle, and they contain much valuable timber. There are two towns, but the principal town is in ruins, and the other a mere village.

Sonepur (area, 1,467 sq. m.; pop. 66,015).—A flat, well cultivated district, on the Mahanuddy. The heat is intense.

Sucti (area, 268 sq. m.; pop. 12,060).

Cuttack Mehals.

A small group of native states, behind the British district of Cuttack, and between that district and Chota Nagpur. They consist of—

Autmalik (area, 648 sq. m.; pop. 29,160).

Boad (1,377 sq. m.; pop. 61,965).—The town is on the Mahanuddy, which is navigable to this point from the sea 190 miles. It was formerly more important.

Duspulla (162 sq. m.; pop. 7,290).—The celebrated car of Juggernauth is made of the sal tree, which grows to enormous size in the forests of this state, and supplies the timber.

Kunjerry (*Keunjur*) (5,022 sq. m.; pop. 225,990).—This raj was for a time under British rule.

Mohurbunge (2,025 sq. m.; pop. 91,125).

Nilgur, Nyagur, Sokinda, Tachirrya, and some other very small states, making up a total area of 6,834 square miles, and a population of 346,275.

CHAPTER V.

WESTERN INDIA.

Boundaries and Subdivisions — General Account — BOMBAY — 1. *Bombay Provinces, Northern Division:* — Ahmedabad District—Bombay and Colaba District—Broach District—Candeish District—Kaira District—Surat District—Tanna, or Northern Concan District. *Southern Division:* — Ahmednuggur District—Belgaum District—Canara (North) District—Darwa District—Kuladgi District—Poona District—Rutnagerry District—Sattara District—Sholapur District. 2. *Sind.* 3. *Native States North of Bombay:* Cutch—Guzerat States—Kyrpur—Rewa Caunta. 4. *Native States South of Bombay:* Jinjira—Kolapur—Sattara Jaghires — Sawunt Warri — Southern Mahratta Jaghires. 5. *Portuguese Possessions.* —Daman—Diu—Goa. 6. *Aden.*

Boundaries and Subdivisions.—That part of India which lies to the west of the Bengal Presidency, Central India, and the northern part of Mysore in Southern India, comprises the whole of the Presidency of Bombay and a number of native states. The total area thus included is 206,423 square miles, and the population upwards of 20,000,000.* It extends for nearly 1,000 miles in length, from the southern extremity of the Punjab in the north, in latitude 28° 45′, to the northern extremity of South Canara district on the Malabar coast, in latitude about 14°. It is everywhere narrow, the widest part in the peninsula of Guzerat not exceeding 350 miles. It has a very extended line of coast on the Arabian Sea, reaching from Kurrachi in Sind, to the Malabar coast, broken by

* The total area of British territory in this Presidency, according to a careful comparison of the most recent returns, is 126,648 square miles, containing a population of nearly 13,000,000.

12

two deep inlets, the Gulf of Cutch and the Gulf of Cambay. It is conveniently subdivided into five principal groups of territories of very unequal magnitude and importance:—

	Area in square miles.
1. Northern and Southern Provinces of Bombay (British) . . .	72,245
2. Sind (British)	54,403
3. Native states North of Bombay 4. Native states South of Bombay	78,608
5. Portuguese possessions . . .	1,167

General Account.—Western India is crossed by the Lower Indus, the Myhi river, the Nerbudda, and the Tapti, besides a considerable number of small streams proceeding from the Western Ghats. It is crossed, or rather entered, by the western extremity of the Vindya range and the spurs that extend from that chain northwards, and also by the Satpura range, between the Nerbudda and the Tapti. But its chief mountain chain is that of the Western Ghats, which ranges parallel to the coast, under various local names, from the Tapti southwards, and at a distance varying in different places, but averaging about fifty miles. Much of the culminating ridge of the chain is within the district. The climate varies considerably, as might be expected from the great extent in latitude and the difference of level above the sea. In Sind it is sultry and very dry. Further to the south the coast also is sultry; but the greatest heats are not so excessive as in many other parts of India. The mean temperature is there about 80°. The rainfall also, further south, is heavy, averaging at Bombay between seventy and eighty inches, but the extremes are very wide. The fall in one year at Mahabaleshwar has been recorded at 248 inches; in the Poona collectorate, on the other hand, the average fall is not twenty inches, and at Kurrachi it is only about seven. The natural productions are cotton, rice, millet, barley, and grain. These are grown everywhere. Sugar-cane and coffee are grown in some places, wheat and the potato in Guzerat, indigo in Candeish. The forests contain valuable timber, gums, drugs, and dyes. Cocoanut-palms form a fringe along the coast, and plantains, mangoes, and the common Indian fruits are grown in abundance. The animals are those of other parts of India.

Of the population the majority is Mahratta; but there are great numbers of Jains, Guzeratis, Bheels, Parsees, and Jews, besides Europeans. The prevailing religion is Hindu; but Mahomedans are numerous. The languages spoken in the south are Mahratta and Canarese. Guzerati is spoken in the north. Portuguese, Arabic, Urdu, Persian, and English are all employed.

Agriculture is the great employment of the bulk of the people. Sugar, indigo, and silk are made in many places. At Poona paper is made. Pottery, of a common kind, is manufactured everywhere. Weaving is carried on to a trifling extent; but there are no manufactures on a large or important scale. There are admirable means of communication, the roads being well designed and well constructed, giving to all important places a ready outlet for manufactures and produce. In addition to these, the lines of railroad executed and now in operation have already opened out the country, and are calculated to do so much more. The railway system of the Presidency will be found described in pp. 39, 40.

Of mineral and vegetable productions, it may be sufficient here to mention that coal is obtained in Cutch; hemp, jute, and other fibres from Sind, Surat and Sattara, besides the coast of the north and south Concan; indigo from Sind, the eastern part of Guzerat, and near Surat, and cotton from Sind, Cutch and Guzerat, and the country between the Myhi and the Nerbudda. There are forests between the Nerbudda and the Tapti, and also at Nassic, in the district of Ahmednuggur.

The Presidency of Bombay is divided into three principal divisions—Northern, Southern, and Sind. Each is under a Commissioner, and they are subdivided into twenty districts, each governed by a Magistrate Collector.

Bombay, the capital of the presidency, is situated in N. lat. 18° 53′ and E. long. 72° 48′, at the south-eastern extremity of the island of the same name, which is connected with another island, *Salsette*, by an artificial causeway and bridge. The island of Bombay is about eight miles long, and is composed of two ranges of rock of unequal length, with an intervening hollow or valley about three miles wide, which has been converted from an unwholesome swamp into a healthy residence. Besides being united to Salsette (which is an island eighteen miles in length and ten miles wide), Bombay island is also joined by a causeway to Old Woman's Island to the north, and Colaba to the south. This long line of natural breakwater in front of the coast of the mainland has produced a land-locked harbour of fifty square miles, which is increased to eighty by including a shallow bay to the north of Salsette. The east side of this harbour is rendered picturesque by several islands, one of which is *Elephanta*, so called from a rude figure of an elephant in black marble cut out of the rock. In this island are celebrated and magnificent cavern-temples, of the most singular and magnificent proportions, excavated out of the rock. The chief temple consists of three sanctuaries or shrines, the largest being about 130 feet square. The figures, symbols, and sculptures relate to the mysteries of Hindu mythology.

The town of Bombay is divided into two parts—the European and the Native. In the former the houses are well built, lofty, and handsome, with many churches, mosques, and other fine buildings. In the latter the houses are small, the streets narrow and crowded. Externally the appearance is attractive, and indicates great wealth, the public buildings of the Government and private establishments being prominent and handsome. The hospital and caravanserai called Daramsala, due, to a great extent, to the princely liberality of the well-known Parsee, Sir Jamsetjee Jejeebhoy, are especially remarkable. The

Parsees generally are the most progressive and most enlightened people of Western India.

The harbour of Bombay is the largest and safest in India, and has every requisite for a great seaport. It is easy of access, affords excellent anchorage, is sheltered, and admirably adapted for docks and shipbuilding. Ships of the largest size are here constructed. The commerce is of great extent and importance.

The population of Bombay exceeds 800,000, and is of the most varied character, including, besides the English, Parsees, Hindus, Mahomedans, Jews, Portuguese, Chinese, and Malays; and, indeed, people from every part of Asia, all speaking their native languages and wearing their national dress. Of these, the Hindus are more than half, and the Mussulmans and Parsees are nearly equal in numbers. All the rest together hardly amount to more than one-fifteenth of the population. The city is not unhealthy, the death-rate differing little from that of London.

Bombay is distant W. from Calcutta, 1,040 miles; NW. from Madras, 645; SW. from Delhi, 730; NW. from Poona, 75.

1. Bombay Provinces.

Northern Division.

Ahmedabad (lat. 21° 22′—23° 30′; long. 71° 26′—72° 50′; area, 4,356 sq. m.; pop. 700,000).—This is one of the two most northerly of the districts of the Bombay Provinces. It is level, the hilly tracts of the Meywa terminating at its extreme north, from which the country, at first undulating, soon subsides into flat plains. It is bounded on the east by the Sabur Mutti river, which separates it from the collectorate of Kaira, but is unnavigable. A part of the district between the head of the Gulf of Cambay and the Runn of Cutch is liable to be flooded. The hills in the north yield a fine building-stone, but road material is very difficult to find, and thus the roads are heavy and in bad condition.

Railways have now rendered this comparatively unimportant. The soil is very fertile, and produces wheat, rice, millet, grain, sugar-cane, and cotton. The water for irrigation is drawn from wells. The inhabitants are well off and well clothed, and occupy comfortable brick-built houses.

Ahmedabad (pop. 130,000) is the principal town. It is on the Sabur Mutti, at the northern extremity of the district. It was founded in 1412, and at the commencement of the seventeenth century was one of the finest cities in India, the walls being still of vast extent, and strong. Their circuit, however, is now less than six miles, and the place is much decayed, although numerous ruined mosques, palaces, mausolea, aqueducts, and fountains attest its ancient glories. The great mosque (Jumma Musjid), the mosque of Sujat Khan, and the ivory mosque (built of white marble, and lined with ivory inlaid with gems) are among the most magnificent structures of their kind in India. Near the city are beautiful gardens, and a small artificial lake, with an island on which is a summer palace. There is a good supply of water to the city, there are good schools, and an English church and chaplain. Distance N. from Bombay, 290 miles; W. from Calcutta, 1,020; SW. from Delhi, 190. *Dolera*, a town near the head of the Gulf of Cambay, has lately become important, and now numbers several thousand inhabitants. A tramway has been laid to the coast. *Gogo*, on the west coast of the Gulf of Cambay, has good anchorage, and is safe during the south-west monsoon. Near it is *Perim*, a small island, in which have been found a number of tertiary fossils of the sub-Himalayan period. The best Lascars in India are from this place. Water and firewood are scarce. Distance N. from Bombay, 190 miles.

Bombay and Colaba (lat. 18° 26′—18° 48′; long. 72° 55′—73° 12′; area 18 sq. m.; pop. about 60,000).—A district including the islands and territory around Bombay and an island on the Concan coast, a little south of Bombay. The country is rich in teak forests and other timber. The island of Colaba is well situated for shelter, and has been a resort for pirates. *Bombay*, the only important town, has been already described (*see* p. 180).

Broach (lat. 21° 22′—22° 11′; long. 72° 30′—73° 10′; area, 1,319 sq. m.; pop. 300,000).—A small district on the east shore of the Gulf of Cambay, bounded on the north by the Myhi, and on the south by the Tapti, and crossed by the embouchure of the Nerbudda and by the River Dadur. It is thus well watered, and includes the delta of the Nerbudda, but is not unhealthy, and has a moderate rainfall. The country is level, and the roads tolerable. Cotton is an important crop. The town of *Broach* is on the right bank of the Nerbudda, about thirty miles above the mouth, and has been a large and flourishing place. It had much decayed about twenty years ago, but has since revived, and now has 80,000 inhabitants, and exports large quantities of cotton. Distance N. from Bombay, 190 miles.

Candeish (lat. 20° 10′—21° 58′; long. 73° 37′—76° 20′; area, 9,311 sq. m.; pop. 800,000).—A district in the interior, north of Ahmednuggur. It is an extensive basin, traversed from east to west by the Tapti, and flanked on the north side by the Satpura range, and on the west by the Western Ghats. The lower part of the district is fertile, but much of it is now covered by jungle. The roads are good, and a great deal has been done to improve the condition of the people. A large part of the population consists of Bheels, who were formerly very troublesome, but are now reduced to order. *Dulia* is a considerable town on the main line of road from Bombay to Agra, 181 miles north-east of Bombay. *Mulligaum* is another large town nearer Bombay, and not far from the Great Indian Peninsula Railway. Both are on tributaries of the Tapti.

Kaira (lat. 22° 12′—23° 33′; long. 72° 30′—73° 27′; area, 1,869 sq. m.; pop. 600,000).—The northernmost district of these provinces. It is traversed by the Bombay and Baroda Railway. The country is flat, and destitute of canals and navigable rivers. It produces the usual grains, tobacco, indigo, cotton, and opium. The principal town, *Kaira*, is of considerable size, and situated on a small stream in the midst of a fertile and beautiful country. The town is surrounded by a wall, and contains a curious Jain temple, partly subterranean. The climate is hot and unhealthy. Distance north from Bombay, 265 miles.

Surat (lat. 20° 15′—21° 11′; long. 72° 45′—73° 24′; area, 1,629 sq. m.; pop. 500,000).—A small district on the coast south of the Tapti, resembling Broach, which it adjoins. The roads are sandy, and the country flat. The town of *Surat*, on the Tapti, at a distance of twenty miles from its mouth, is large, and has 130,000 inhabitants. Its trade was formerly much more important than at present. The town is ugly and uninteresting, and the river of little use, as it is fordable when the tide is out, and has a troublesome bar. Distance N. from Bombay, 150 miles.

Tanna, or Northern Concan (lat. 17° 56′—20° 20′; long. 72° 42′—73° 48′; area, 5,795 sq. m.; pop. 880,000).—A strip of territory on the mainland, commencing east of Bombay and Salsette Island, and extending for some distance northward. It includes hilly tracts, intersected by ravines, which are excessively rugged, and covered with jungle infested by wild beasts. On the hills are forests of teak and other timber. There is no river of importance, and no harbours sufficient to shelter ships, but there are many creeks on the coast, where pirates formerly found refuge. The crops are not of much importance, though on the low lands they include those common in similar localities in other districts. *Tanna*, the chief town, is a flourishing place of 10,000 inhabitants. It is on the line of railway from Bombay to Nagpur, and is only twenty-four miles from the capital.

Southern Division.

Ahmednuggur (lat. 18° 16′—20° 30′; long. 73° 29′—75° 37′; area, 9,931 sq. m.; pop. 1,000,000).—A district extending for a considerable distance parallel to the coast behind Bombay, and within the line of the Western Ghats, which form part of its western boundary. Several spurs are thrown out eastward from the ridge, and between them are table-lands slightly inclining towards the south-east. These are beautiful, and highly cultivated. The country is traversed by good roads, which form the communication from the Nizam's dominions to the coast. The produce is chiefly grain. The manufactures include silk and coarse cotton cloths.

Ahmednuggur, the principal town, is naturally defended by a living wall of cactus (prickly pear), about twenty feet high, and very thick. The town is large and increasing. It has a *choultry*, a place for the accommodation of travellers, large enough to hold 250 persons. Distance E. from Bombay, 122 miles; SW. from Calcutta, 930 miles. *Nassic*, formerly *Panchavati*, a sacred city among the Hindus, and more revered even than Benares. Notwithstanding this, it contains many Buddhist excavations and remains. Nassic is a very important and improving town, and from its position on the main line of the Great Indian Peninsula Railway, its salubrity, and security from sudden attack, it has been suggested as the fittest place to adopt as the seat of the Supreme Government of India. It is about 100 miles from Bombay.

Belgaum (lat. 15° 23′—16° 39′; long. 74° 2′—76° 23′; area, 5,405 sq. m.; pop. about 1,000,000).—A district in the southern part of the presidency, within the Ghats, and adjoining the Portuguese territory of Goa and some small independent states. The town of *Belgaum* is small, but has been recently much improved. It has important educational institutions. It is situated on the plain east of the Ghats, 2,500 feet above the sea. Distance S. from Bombay, 246 miles.

Canara, North (lat. 13° 35′—15° 30′; long. 74° 9′—75° 10′; area, 4,300 sq. m.; pop. 50,000).—The most southerly of the Bombay collectorates, extending southwards on the coast from Goa, and having a coast-line of nearly 80 miles. It is one of the most fertile districts in India; but from one end to the other of its long line of coast there is not a safe station for vessels of even moderate size, although there are numerous creeks and inlets. The plains between the foot of the Ghats and the sea are studded with cocoanut-palms and rice-fields. The hill-slopes produce cardamoms, pepper, and areca-nuts; and the summits of the Ghats are crowned with dense forests of teak and other valuable woods. The population is chiefly on the plains and along the coast, the hilly districts being unhealthy. There are several small streams running from the mountains to the sea, two of them forming magnificent falls, one of them 880 feet. The towns in North Canara are few and small. *Honawar* is the principal, and was once rich and beautiful, though now in ruins. *Cumta* was also once of some note, and is now a port for the shipment of cotton from the southern Mahratta country to Bombay. *Carwar*, on the coast, is well adapted for a harbour.

Darwa (lat. 14° 16′—15° 50′; long. 74° 50′—76°; area, 3,837 sq. m.; pop. 1,000,000).—One of the most southerly districts of the Bombay Presidency adjoining North Canara. A great part of it consists of extensive plains among the Western Ghats, but inclining to the south-west, and traversed by streams that enter the Arabian Sea. Many parts are fertile and grow cotton; and there are good roads for the conveyance of produce. The town of *Darwar* is of no great importance. *Hubly*, though ill built, is thriving, and has much trade. It is one of the principal cotton marts of the southern Mahratta country. Distance SE. from Bombay, 290 miles.

Kuladgi (lat. 15° 50′—16° 35′; long. 75° 20′—76° 20′; area and population undetermined).—A district beyond Sattara, and formerly part of that collectorate. It is on the plateau, and borders on the Nizam's dominions, being watered by tributaries of the Kistna. *Kuladgi* is the chief town. It is on a small tributary of the Kistna. *Bijapur* is a ruined town, once the capital of a powerful kingdom, with lofty walls, presenting from the outside the appearance of a flourishing city; but within all is solitude, silence, and desolation. The chief objects are the mausoleum of Mahommed Ali Shah, the great mosque, and the tomb of Ibrahim Adil Shah. Both the former are grand edifices, and the latter is exceedingly graceful. Distance from Bombay, 245 miles.

Poona (lat. 17° 53′—19° 26′; long. 73° 20′—75° 10′; area, 5,298 sq. m.; pop. 750,000).—A large district in the interior, immediately to the east of Bombay. It is an elevated table-land, with a few hills rising above it; and some small spurs of the Western Ghats, part of whose culminating ridge forms its western boundary. Several streams cross it, which ultimately enter the Beema, a tributary of the Kistna. There are good roads across the district. The climate is very dry, but not unhealthy; and the vegetation is adapted to the climate. There are few trees, but by irrigation good crops of rice, maize, millet, cotton, sugar-cane, and potatoes are obtained. The inhabitants are Mahrattas. The chief town is *Poona*, on the Muta, in a treeless plain 2,000 feet above the sea, and overlooked by the Ghats, which rise 1,000 feet above the plain. It is a station on the Great Indian Peninsula Railway. It has recently been greatly improved, and almost rebuilt. Its population is now 100,000, but is said to have been much more formerly.

Poona is an important military station. Distance SE. from Bombay, 74 miles.

Rutnagerry, or **Concan** (lat. 15° 44′—18° 6′; long. 73° 6′—73° 58′; area, 3,964 sq. m.).—This district, which is also called the Southern Concan, is a long strip of land between the coast of the Arabian Sea, south of Bombay, and the Western Ghats. The quantity of arable land is small, as the rise from the coast to the mountain ridge is rapid, and the soil rocky, but on the whole, cultivation is good, and the people are well cared for. Rice and grain are the chief products. The chief town, *Rutnagerry,* is on the coast. *Viziadroog,* or *Geria,* is another town. Neither is important, but Geria has an excellent harbour. It is 170 miles S. of Bombay. *Vingorla* is another small town in a sheltered bay. It is 215 miles S. of Bombay.

Sattara (lat. 16° 22′—18° 32′; long. 73° 24′—76° 25′; area, 10,222 sq. m.; pop. about 1,000,000 *).—A large district on the eastern side of the Ghats, divided into two parts by a ridge nearly parallel to the chain of the Ghats, extending for sixty miles to the Kistna.. Near the Ghats the surface is rugged, and the climate wet, the rainfall in some places being nearly 300 inches. On the eastern side, however, the rains are light and uncertain, and the rainfall not more than twenty-four inches. The climate is healthy. The productions are cereals, pulses, fruits, coffee, flax, cotton sugar, opium, and tobacco. Within the district, and beyond it to the west, are possessions of chieftains who hold the land under the English by a kind of feudal tenure. *Sattara,* the chief town, is immediately below a remarkably strong hill fort, on a detached summit, 1,100 yards long, and 500 yards across. The scarp is about forty] feet of perpen-

* The area and population include those of the district of Kuladgi, recently separated.

dicular rock, above which is a stone wall. It has been frequently taken. It lies SE. from Bombay 115 miles. *Punderpur*, on the Beema, is a large and populous town, with a temple of peculiar sanctity.

Mahabaleshwar is a small town and sanatorium about 4,500 feet above the sea, on a rugged and undulating table-land of the Western Ghats, whose western buttresses rise abruptly from the adjacent Concan. The greatest breadth of this plateau is at the northern end, and is about fifteen miles; but its extent, taken diagonally, is seventeen miles. It is connected with the Ghats further south by a narrow isthmus. The district is remarkable for the violence of the monsoons and the abundance of rain. Rain falls on 227 days to the extent of 229 inches. The water, however, runs off very quickly; and there are no marshes. The winter season is delicious. The town is about 4,500 feet above the sea. Distance NW. from Sattara, 30 miles.

Sholapur (lat. 16° 10′—18° 34′; long. 75°—76° 28′; area, 4,991 sq. m.; pop. 700,000).—A district on the left or eastern bank of the Beema river, reaching to the Kistna, which forms its southern frontier. It is an undulating surface, presenting a selection of upland and valley, well watered, but devoid of trees. The climate is dry and healthy, the rainfall being exceedingly small, but several streams cross the district. Cotton is the staple product, and is sent in enormous quantities to Bombay. The roads were bad, but the railway from Bombay to Madras is now open to the town of *Sholapur*, which is in the eastern part of the collectorate. It is an improving place. Distance SE. from Bombay, 220 miles. *Barsi* is a depôt for cotton.

2. Sind.

The province of Sind lies between lat. 23° 35′ and 28° 30′; long. 66° 20′—71°. Its area is 54,403 sq. m.; and its population about 1,800,000. It is detached from the other provinces of Bombay Presidency by Cutch and

Guzerat, and the greater part of it lies between Rajputana and Beluchistan, with a frontage of about 130 miles on the Arabian Sea. The north-eastern portion consists partly of the desert called the Thur, bordering on the Great Runn of Cutch. There are four districts—Kurrachi (area 19,240 sq. m.; pop. 330,000), Hydrabad (area 23,974 sq. m.; pop. 760,000), Shikarpur (area 9,042 sq. m.; pop. 650,000), and the Frontier of Upper Sind (area 2,147 sq. m.; pop. 50,000.)

This large and important province comprises the lower course and the delta of the Indus, and forms the westerly extremity of the British dominions in India. Its name is synonymous with that of the river, which is called indifferently the Sind or the Indus. On the melting of the snows this great stream rushes furiously down from the distant Himalaya to the sea, breaking down banks and whirling along with it trees and every other object coming within its influence. In winter it is calm and sluggish. The width of the modern delta is 150 miles, from the Kori, which separates it from the Runn of Cutch, to Cape Monze, fifteen miles beyond Kurrachi. A large part of it is submerged at spring tides. Along this great length of coast Kurrachi is the only safe harbour. The delta may be said to commence at Tatta, about fifty miles from the extremity, and its surface is about 3,000 square miles. It is almost level, and nearly destitute of timber. The soil is alluvial, except in some rocky portions of small extent.

Beyond the delta on each side are other alluvial tracts of ancient delta, having an exceedingly rich soil, whose fertility is exceeded in no country on the earth. Beyond this, again, is a considerable breadth of land, also very fertile when irrigated, consisting of the *doab*, or tongue-shaped interval between the Narra branch and the Indus, whose width is seventy or eighty miles, and which continues for 300 miles. In this part, however, are two limestone ranges, a hundred feet above the sea. Although very rich, much of the soil in the lower part of the Indus valley yields salt on evaporation, and for this, as well as other reasons, irrigation is necessary for profitable cultivation. The actual soil is generally a stiff greasy clay, occasionally intermixed with sand.

The climate of Sind is remarkably dry and sultry. The recorded rainfall at Kurrachi and Hydrabad is so small as to be almost nominal, and further north, at Larkana, three years have elapsed without rain. Occasionally, no doubt, there are heavy showers and severe storms, even in the rainless country, but the country is so situated as to escape the influence of both monsoons. The range of the thermometer is very wide. In summer, at Sukkur, the temperature is 102°, and further north it is even greater. In winter there are frosts.

The soil yields two crops in the year when watered. The first crop is sown in spring and reaped in autumn, and consists of rice, maize, sugar, indigo, and cotton; the crop sown in winter and reaped in spring consists of wheat, barley, millet, oil-seeds, hemp, and tobacco. Fruits of many kinds are grown, and gigantic grasses furnish excellent material for ropes and thatch. There is a rich variety of animal life, including camels and buffaloes, tigers and other feline animals, wild asses, wild hogs, and deer.

The Sindians are a mixed race of Jats and Beluchis, and are partly Hindus, partly Mahomedans, in religion. The men generally are a fine people, and the women are proverbially beautiful. The fishermen are particularly handsome. But in every town, and on the banks of the Indus, there are swarms of lazy worthless beggars, or fakirs, who extort alms from the poor cultivators of the soil, under the pretence of religion.

The language of Sind, called Sindi, is a dialect of Sanskrit. West of the Indus Beluchi is spoken. Persian is used by the educated classes.

The manufactures include silk and cotton, cloth, paper, leather, swords, and fire-arms. Earthenware is made in all the towns, and gunpowder in most. The natives excel as weavers, turners, and dyers, and their wooden lacqured work is well known. The trade of the country is rapidly increasing. Since its annexation, the commerce has been set free from absurd restrictions; roads and canals have been made, waste lands brought under cultivation, and schools and other educational institutions organized. The effect is already very manifest.

There are numerous towns in Sind, some of them of considerable importance, and having large and busy populations. The following are the most important:—

Hydrabad, the capital, is situated four miles east of the Indus, on a low rocky range called the Gunja Hills, and on an island between the main stream of the Indus and the Fulailee branch. It has a fortress, once regarded as strong, but not now defensible, and walls which, with their bastions, are more picturesque than useful. It is a poor, badly built town, with one principal street, and an extensive bazaar. There are some handsome tombs in the cemetery. Distance NW. from Bombay, 751 miles. *Kotri*, opposite Hydrabad, is now connected with Kurrachi by rail, and with Multan in the Punjab by steam. *Meani*, the site of a celebrated battle won by Sir Charles Napier, with 3,000 men over the Ameers of Sind with 22,000 followers, is six miles north of Hydrabad.

Kurrachi.—This place, the chief port of Sind, is situated almost at the western extremity of the country, and close to the extreme boundary of British India. It is near the base of a low hill range, on a level space extending to the sea, well sheltered by a rocky headland, which projects south-eastwards from the mainland, leaving a space of about two miles, where a good natural harbour is formed, containing, however, both sandbanks and rocks, and having a bar at the entrance. The harbour is spacious, and though the town is three miles from the landing-place when the tide is out, it has been rendered easy of access. Kurrachi is a place of very great commercial, political, and military importance, and its trade is increasing steadily and rapidly. It is the terminus of the Sind Railway, already open to Hydrabad. Between the two towns is *Tatta*, formerly wealthy and important, but now miserable, unhealthy, and chiefly inhabited by beggars. It is close to the head of the modern delta of the Indus. Near it are remains of ancient cities, with some fine specimens of building, and a cemetery six square miles in extent.

Larkana is a town situated in one of the most fertile tracts in Sind, about seven miles from the Indus. It is one of the chief grain marts of India, and has a good bazaar, some silk and cotton manufactures, and about 12,000 inhabitants. It is one of a group of three towns in the northern extremity of Sind. *Rori*, or *Lohuri* (population 8,000), is on the left bank of the Indus, built on a rocky eminence of limestone, forty feet above the river. It has a striking appearance from without, as the houses are lofty; but they are badly built and ruinous. The freshets of the Indus at Rori rise sixteen feet above the lowest level of the stream, and thus render the position of the town very convenient. The streets, however, are very narrow, and the air is close and unwholesome. The number of mosques is enormous. In the largest there is preserved a single hair in amber, supposed to be a hair of the beard of Mahomed, kept in a richly jewelled case. *Bukkur*, close by, is a huge fortress, built on a rocky island in the channel of the Indus. *Shikarpur* is one of the most important commercial towns in Sind, with 30,000 inhabitants. It lies twenty miles west of the Indus. It has a great transit trade, and is the resort of Hindoo merchants, who have commercial relations all over the East. It is at the junction of routes in every direction. *Sehwan* is a small mud-built town, near which is a tomb much visited by pilgrims. *Sukkur* is a decayed town on the right bank of the Indus. It is very picturesque, and of some commercial importance, but is only half the size of Rori. *Vikkur* is a town on one of the branches of the Indus, within the delta, 60 miles SE. of Kurrachi.

3. Native States North of Bombay.

Cutch (lat. 22° 47′—24° 40′; long. 68° 26′—71° 45′; area, 6,500 sq. m.; pop. 409,522).—This singular tract of almost detached land, about 205 miles from east to west, and 110 miles across, is interposed between the desert tract in the south of Rajputana and the sea, and forms a connecting link between Guzerat and Sind. It is intersected by two hill-ranges of moderate elevation. Both indicate volcanic activity. In the valley between, and in the plain to the south, there are large fertile tracts. On the northern side of the hills there is also a broad belt of luxuriant pasturage. The country is characterized by a deficiency of water, although, during the rainy season, numerous torrents descend from the hills, and sometimes cover the low ground. The sub-soil is porous, and no supply can be secured in reservoirs, but wells are numerous, and yield a good supply of excellent water. Among minerals, coal, iron ore, and alum are obtained. Rice, millet, sugar-cane, cotton, and some fruits are the chief vegetable products; but during a great part of the year, large tracts of country exhibit nothing but a rocky and sandy waste, and the produce of the cultivated lands is not sufficient to support the scanty population. The province abounds in game and wild beasts, and it possesses a peculiar breed of horses, besides a beautiful species of wild ass in large herds. The climate is healthy, temperate, and agreeable, except during the three hot months, when the heat is excessive. In winter, on the other hand, the cold is severe.

The land of Cutch is separated from the main land of the Indian Peninsula to the north and east by salt marshes of enormous dimensions, called "the Runns of Cutch." That to the north (the Great Runn), is about 190 miles in length, varying in breadth from two to ninety miles, and has a total area of upwards of 7,000 square miles. The smaller is a triangular space, about seventy miles a side, and its area is 1,600 square miles. These vast level spaces are alternately swamps, deserts, and lakes. In the dry season, they are sandy wastes, interspersed with wide sheets of shallow pools of salt water, ridges of sand, and patches of tamarisk. In the rainy season, they are covered about knee-deep with water. In this respect they resemble the deltas of some large rivers. There are several islands within the boundaries of each, and peninsulas enter them from the main land. The bottom is slimy, hard, dry, and sandy, and clay is rare. The quantity of salt is so great, that the surface is often encrusted an inch deep, and crystalline lumps of salt as large as a man's fist may be picked up. They are occasionally flooded by sea-water blown into them. At other times they are entirely inundated by rain-water and swollen streams. At all times fresh water is scarce, except in the rocky islands. The wild ass roams throughout. The phenomenon of the mirage (called *sirab*) is very strongly exhibited, and magnifies objects very highly, so that patches of shrubs resemble forests, and the wild asses, the largest animals, appear as large as elephants. During the dry season, the reflection of the sun from the glazed saline surface resembles the sparkle of water. Flies are so numerous that it is almost impossible to breathe without swallowing some; and even if they do not bite, it is difficult to force a horse through their swarms. Several roads, passable by vehicles, cross the Great Runn in the narrowest portions; but to cross it during day time, in the dry season, is almost impossible. Besides the wild ass, which is peculiar to the Runns, there are apes, and porcupines, and vast flocks of birds. On the subsidence of the water after the rainy season, multitudes of dead prawns and fish are strewn over the surface, and become very offensive. During an earthquake in 1819, a mound of earth, many miles in extent, was uplifted, and large tracts of land submerged.

The inhabitants of Cutch are partly Hindus and partly Mahomedans. The ruling class are *Jarejas*, a branch of the Rajput tribes. They are a singularly fine race of people, robust and warlike, but dissipated, proud, and cruel. The mariners are fearless and enterprising, and the best pilots in India.

The government of Cutch is peculiar. There are about 200 chiefs, who exercise unlimited authority within their respective domains, and form a kind of brotherhood or council. Over these is a superior chief or king, called the Rao, who advises with them on all Political Affairs. The English are represented by a political agent appointed from Bombay. The annual revenue of the Rao is about eight lakhs of rupees (£80,000).

Booj is the capital of Cutch. It is situated at the base of a fortified hill near the centre of the district. Viewed from the north, it has an imposing appearance, the number of white buildings, pagodas, and mosques, interspersed with plantations of date-palms, giving an idea of respectability entirely removed on entering the town. The streets are narrow, dirty, and almost impassable, owing to the numerous herds of sacred bulls. The Rao's palace is a large white stone castle, enamelled outside and decorated with beautiful carvings. There is a fort. *Luckput*, seventy-one miles W. of Booj, is on the SE. bank of the channel which connects the Great Runn with the Arabian Sea. This channel was once a branch of the Indus, but is now a mere creek. It is on elevated ground, and there is a fortress. *Mandavi*, thirty-four miles SW. of Booj, is the principal seaport, and has considerable trade; it is on the Gulf of Cutch.

Guzerat States.—The territory thus named, including the Peninsula of Katiwar and other dominions of the Guicowar and his tributaries, besides a number of independent states, is bounded by Cutch, Rajputana, Central India, the Bombay Province of Candeish, and the Arabian Sea. Its limits are lat. 19° 50′—24° 45′; long. 69°—74° 20′; and it contains about 55,000 square miles, of which area nearly half is comprised within the peninsula. The population exceeds four and a half millions.

This tract is for the most part flat, and yields cotton, rice, wheat, barley, millet, grain, sugar-cane, and fruits, in abundance. Amongst animals, the quadrupeds include lions (a maneless lion), tigers, leopards, wolves, hyænas, and deer. Camels, buffaloes, oxen, and horses, are common; the wild ass is found in the uncultivated tracts, and with it wild cattle resembling the bison. The flamingo, adjutant bird, and many water-fowl are common. The mainland is watered by the Rivers Sabur Mutti, Bunass, Myhi, Nerbudda, and Tapti. The Western Ghats constitute the eastern boundary of the district, but they are not lofty. There are hardly any ordinary roads, but the railway is now available to Baroda, the capital. Of the people, the ruling tribes are Mahrattas. Rajputs, Jains, and Bramins abound, the latter being the landed proprietors. Mussulmen, Boras, and Parsees are found in the towns; and Coolies, Konds, Katties, and Bheels in the country. There are also two singular classes of people attached to the Rajputs, called Bats and Charuns, both of whom constitute themselves safeguards against attack, as being sacred and of celestial origin. (*See* Rajputana.)

At *Champaneer* is a hill fort and a remarkable temple. *Dubbooe* is an ancient town, with interesting Braminical remains.

The following are the states included in Guzerat. Most of them are very small. The tribute is collected by the British, who pay the Guicowar (to whom most of the petty states are subject) the share that belongs to him. "The Guicowar" is the title of the Mahratta chief of Katiwar and other dominions of the district of Guzerat, who is also the suzerain of the petty states.

Balasinore (lat. 22° 53′—23° 17′; long. 73° 17′—73° 40′; area, 258 sq. m.; pop. 19,092.)—A small state on the Myhi, near the district of Kaira, held by a Nawab. There is a thriving town, well supplied, and surrounded by a wall. Distance N. from Bombay, 280 miles.

Bansda (lat. 20° 35′—21°; long. 73° 8′—73° 28′; area,

325 sq. m.; pop. 19,000).—A small state south of the Tapti, and east of Surat.

Baroda (lat. 21° 46′—22° 51′; long. 72° 50′—73° 48′; area, 4,399 sq. m.; pop. 1,710,404).—An important state, reaching to the northern extremity of the Gulf of Cambay, and bounded by the Rivers Nerbudda and Myhi. The western extremity of the Vindya range, expanding into the Barria Hills, reaches the northern extremity of the state, and gives rise to streams which water it. This state is the dominion of the Guicowar, or Guikwar, that being the title assumed by the Raja. *Baroda*, the capital, is near the river Biswamintri. It is a place of importance, connected by railway with Bombay, from which it is distant 231 miles. The city is large and well built, with a population of 140,000. It is crossed by two spacious streets, the market-place, in the centre of the town, having a square pavilion. The houses are lofty, with tiled roofs. The palace of the Guicowar is a plain building, with projecting wooden galleries.

Baubier (area, 120 sq. m.; pop. 500).

Cambay (lat. 22° 9′—22° 41′; long. 72° 20′—73° 5′; area, 500 sq. m.; pop. 37,000).—A small state at the head of the Gulf of Cambay, governed by a Nawab. The ancient city of Cambay is on the north side of the estuary of the Myhi river. It is now small and unimportant, but was formerly a place of great trade. It was long celebrated for its manufactures of chintz, silk, and gold stuffs; and though these have failed, it still enjoys a celebrity for manufactured articles of agate, cornelian, onyx, &c., obtained from a small deposit near the Rajpipla Hills, on the banks of the Nerbudda. Before being worked, these stones are exposed to the heat of the sun for two years or longer. Distance N. from Bombay, 230 miles.

Charcut (area, 80 sq. m.; pop. 2,500).

Chowra (lat. 23° 35′—23° 56′; long. 70° 53′—71° 11′; area, 225 sq. m.; pop. 2,500).—A very small district, forming, during the rainy season, an island in the Runn of

Cutch. The country is flat, and much salt is found. This little state pays no tribute, and is perfectly independent.

Daung Rajas (lat. 20° 22′—21° 5′; long. 73° 28′—73° 52′; area, 950 sq. m.; pop. 70,300).—A group of petty states between Surat and Candeish, tributary to a chief styled the Rajah of Daung. The country abounds with teak forests.

Deodar (area, 80 sq. m.; pop. 2,000).

Durrumpur (lat. 20° 5′—20° 24′; long. 72° 55′—73° 35′; area, 225 sq. m.; pop. 15,000).—A small state east of the southern extremity of Surat, overrun with dense forest, and admitting of little cultivation.

Hursul (see *Peint*).

Jowar (area, 300 sq. m.; pop. 8,000).

Katiwar (lat. 20° 42′—23° 10′; long. 69° 5′—72° 14′; area, 21,000 sq. m.; pop. 1,500,000).—A province, comprising the peninsula of Guzerat, and including a large part of the dominions of the Guicowar, which, however, extend to the mainland, both north and east. It is divided into ten districts, each subdivided into the separate possessions of a host of Hindu chiefs, some of whom are tributary to the British Government, others to the Guicowar. The surface of the country is undulating, with low ranges of hills in irregular directions. The land in the middle is highest, and here several streams take their rise, some of which enter the Runn of Cutch, others the Gulf of Cutch, the Gulf of Cambay, or the Arabian Sea. There is a rugged part called the Gir, consisting of a succession of ridges and hills covered with forest trees and jungle. The highest ground here is about 1,000 feet above the sea, and there are caverns, ravines, and other fastnesses, very difficult of access. The water is bad, and the climate deadly. Among the wild animals of this part are migratory rats of extraordinary size. In the cultivated part of the country, millet, maize, and wheat, sugar-cane and cotton are grown. The soil is not fertile. *Dwarka* is a town on the western shore of the peninsula, conspicuous from a distance by the great Temple of Kistna, or Dwarkanath, the most celebrated of all the shrines to this god. It is one of the most remarkable Hindu shrines in India. It consists of (1) the *munduff*, or hall of congregation, only 21 feet square internally, but five stories high, surmounted by a dome whose summit is 75 feet from the pavement; (2) the *derachna*, or penetralia; and (3) the *sikra* or spire, consisting of a series of pyramids, terminating at 140 feet from the ground. It is a marvellous construction. In the island of *Beyt* or *Bet* are numerous temples and shrines in honour of Kistna. At *Puttun Somnauth* are also remains of a celebrated temple.

Myhi Caunta (lat. 23° 14′—24° 28′; long. 72° 41′—74° 5′; area, 4,000 sq. m.; pop. 311,046).—A state in the north-eastern extremity of the Guzerat territory, bordering on Rajputana. It has several towns; but, though containing considerable populations, they are not of importance.

Palunpur States (lat. 23° 57′—24° 41′; long. 71° 51′—72° 45′; area, 6,041 sq. m.; pop. 321,645).—A group of petty states in the north of Gujerat, watered by several streams rising in the hill country towards the north-east, and entering the Runn of Cutch. The principal state is *Palunpur* (area 1,850 sq. m.; pop. 130,000). There is only one good road through the district, but that is important. *Palunpur* is a large town of the same name, with 30,000 inhabitants. In *Deesa* another state is a British cantonment.

Peint, with *Hursool* (lat. 20° 1′—20° 27′; long. 73° 10′—73° 40′; area, 750 sq. m.; pop. 55,000).—A small state between the Northern Concan and Ahmednuggur. There are two small towns, *Peint* and *Hursul*.

Radunpur (lat. 23° 26′—23° 58′; long. 71° 28′—72° 3′; area, 850 sq. m.; pop. 45,000).—A small state traversed by the Bunass river, in the north-western corner of Guzerat. It is close to the Runn of Cutch, and yields much salt. The heat is excessive during summer; but from December to April the climate is delightful. The chief town is of considerable size, with a population of 15,000.

Seogaum (area, 64 sq. m.; pop. 4,500).

Suchin (area, 300 sq. m.; pop. 22,000).—A small territory within the collectorate of Surat. The town is ten miles SE. of Surat city.

Thurwarra (area, 48 sq. m.; pop. 800); *Thurrand* (area, 600 sq. m.; pop. 23,000); *Warye* (area, 299 sq. m.; pop. 29,000); *Wow* (area, 324 sq. m.; pop. 71,000).—Petty independent states, on the north-western frontier of Guzerat, bordering on the Runn of Cutch, and north of the Bunass river. None are of importance in any political sense.

Wusravi (lat. 20° 55′—21° 33′; long. 72° 46′—73° 51′; area, 450 sq. m.; pop. 33,300).—This is a native Bheel state, on the Tapti, between Broach and Candeish. It is of small importance.

Kyrpur (lat. 26° 3′—27° 48′; long. 68° 10′—70° 12′; area, 5,000 sq. m.; pop. 105,000).—This is an independent native state, governed by an Ameer. It is almost enclosed by the northern part of Sind, and borders on the state of Bawalpur (*see* p. 125). The country resembles Sind, being alluvial land, with a clay soil, largely mixed with sand. The wells are brackish, and water for irrigation and drinking is procured by a canal from the Indus. The town is little more than a collection of mud hovels, with a few houses of better construction. The palace is among the bazaars. The town is very filthy and unhealthy; the population 15,000. It is 13 miles SW. of Rori.

Rewa Caunta (lat. 21° 23′—23° 33′; long. 73° 3′—74° 18′; area, 8,736 sq. m.; pop. 350,000).—A group of states in the eastern part of Guzerat, comprising:—

Baria (area, 870 sq. m.; pop. 64,380).

Chota Oudeypur, or *Mohun* (area, 3,000 sq. m.; pop. 78,366).

Lunawarra (area, 1,736 sq. m.; pop. 37,000).

Rajpipla (area, 4,500 sq. m.; pop. 122,100).—A flourishing agricultural state, well watered, and having two towns. *Nandode* is the capital. There are celebrated cornelian mines in the state, but they are not now worked. The stones were cut at Cambay.

Soauth, or *Saunte* (area, 900 sq. m.; pop. 31,450).—A country difficult to penetrate. There is a strong fort near the town, crowning a high rocky hill. The state is sometimes called *Soauth Rampur*.

4. Native States South of Bombay.

Jinjira (lat. 18°—18° 32′; long. 72° 55′—73° 15′; area, 324 sq. m.; pop. 71,000).—A small principality, sometimes called the Hubshis, or Hubsies, on the coast of the Arabian Sea, a little south of Bombay, between the Northern and Southern Concan. There is a natural harbour of considerable extent, dividing the state in two parts, and a stream, terminating in an estuary, separates it from the British territory on the north. The harbour is excellent, with four or five fathoms' water everywhere, and shelter from all winds. Off the mainland is the fortified island of Jinjira, on the southern side of the harbour entrance; and the town of *Rajapur* is opposite, on the north side. This place was formerly very important in connection with the pilgrimage to Mecca.

Kolapur (lat. 15° 58′—17° 17′; long. 73° 47′—74° 46′; area, 3,445 sq. m.; pop. 500,000).—A raj, or state, within the Deccan, consisting of a tract, on the eastern side of the Ghats, commencing with the culminating ridge beyond the Southern Concan, and sloping, with a rugged surface, towards the plateau of Kuladgi and Belgaum. There are several streams crossing it, all running towards the east, most of them mountain torrents, tributaries to the Kistna. The Ghats, in the western part of the states, are from 3,000 to 4,000 feet high. The rocks are chiefly volcanic. The population consists of Mahrattas and Ramuses, the latter resembling the Bheel tribes, already described. The government is carried on by a Rajah, subject to British authority. The town of *Kolapur* is a secluded spot, formerly very crowded and unhealthy, but now much improved, and supplied with water. Distance SE. from Bombay, 185 miles.

Sattara Jaghires.—A number of small states, within and adjoining the British district of Sattara. They are *Akulcote* (area, 986 sq. m.; pop. 77,339). *Duflay* (area, 700 sq. m.; pop. 58,794); the most northerly. *Bore* is the town. *Nimbalkur* (area, 400 sq. m.; pop. 47,100). *Phultun* is the town. *Punt Prithi Nidhi* (area, 400 sq. m.; pop. 67,967). *Punt Sucheo* (500 sq. m.; pop. 110,193). *The Waekur.* None of them possess any special interest.

Sawunt Warri (lat. 15° 38′—16° 15′; long. 73° 40′—74° 22′; area, 900 sq. m.; pop. 152,206).—A small state, forming part of the tract called the Concan, or the land between the Ghats and the sea, and situated between the district of the Southern Concan and the Portugese territory of Goa. It approaches, but does not reach, the coast. Its surface, reaching beyond the watershed of the Ghats, is rugged and broken, interspersed with mountains and dense jungle, intersected by small rivers and rivulets, which, at first torrents, gradually become streams of a more regular nature as they approach the coast. The monsoon rains on the higher land and summit of the Ghats are extremely heavy, approaching 300 inches in the year. Tigers, leopards, hyænas, and other wild animals of prey reach even the more fertile tracts, which are covered with luxuriant vegetation. Snakes and other reptiles abound on land, and alligators in the water and swamps. The soil is light and stony, but yields large crops of rice and jowar, besides wheat, gram, and esculent vegetables in the cooler season. There is a military road through the territory, and several native roads. The territories are governed by a Mahratta chief, named Sur Dessayee, subject to British authority. There is a small town (*Sawunt Warri*) twenty-two miles E. by N. of Vingorla.

Southern Mahratta Jaghires.—A group of small states in the South Mahratta country, south of Sattara. They consist of *Bawa*, with the town of *Meeruj*; the *Gorepuray of Mudhole*; *Nepanikur*; and *Putwudun*. The total area is 3,700 square miles, and the population 410,470.

5. Portuguese Possessions.

The Portuguese possessions in India, formerly very extensive and important, are now reduced to the district of Goa and the towns of Daman and Diu, each with small adjacent territories.

Daman is on the coast between Surat and the Northern Concan. It is included within the Presidency of Bombay, and is eighty-two miles from the city of Bombay. It is a small fortified seaport town, on the River Damangunga, which rises in the Concan, about forty miles east. There are roads outside the bar, in which vessels can anchor in eight fathoms' water, and though the river has a bar, there is never less than three fathoms' water at ordinary spring tides. There are docks, and many ships are built here. The surrounding country is fruitful and pleasant, except in the rainy season, when it is inundated. The water is not good, either in the river or from wells. The territory attached to Daman is about ten miles in length from north to south, and five in breadth. Daman was taken by the Portuguese in 1531. Distance N. from Bombay, 101 miles.

Diu is a small seaport town on the peninsula of Katiwar, acquired by the Portuguese in 1515. It is at the eastern extremity of an island seven miles long and less than two miles broad. There is a small well sheltered bay, with good anchorage. The town is well fortified, and is supplied with food from the mainland, the island not being very productive. The place is falling into decay. Distance NW. from Bombay, 170 miles.

Goa (lat. 14° 54′—15° 45′; long. 73° 45′—74° 26′; area, 1,066 sq. m.; pop. 350,000).—The only district of any magnitude remaining to Portugal of her former large possessions in India. It is about forty miles in length, and twenty in breadth, well watered, fertile, well cultivated in most places, and producing rice, pepper, cocoanuts, betel, and salt; but the quantity of rice is not sufficient to feed the inhabitants. The chief city of the district is also *Goa*, which has a fine harbour, almost equal to that of Bombay. The old town contains many good buildings, churches, and monasteries, but is a mass of deserted ruins; and the new town, near the harbour, is low and miserably built. The people are descendants of the Portuguese.

6. Aden.

The territory of Aden, though not properly belonging to India, is now subject to the Bombay Presidency, and is of some importance to the Indian Empire, being a military port of great strength, a depôt for coals, and an entrepot for extensive commerce. It is situated on the high road to India, at the extremity of the Red Sea, on the coast of Arabia Felix, in the Province of Yemen. Lat. 12° 45′ N.; long. 45° 3′ E. Its harbour is the finest in Arabia. The whole place is a peninsula connected with the main land by a sandbank covered at high tides. It is an extinct volcanic crater, one side being open to the sea. The town is surrounded by an amphitheatre of lofty mountains.

CHAPTER VI.

SOUTHERN INDIA.

Boundaries and Subdivisions—General Account — MADRAS — 1. *Madras Collectorates.* Arcot (North and South) Districts—Bellary District — Canara (South) District — Coimbatore District — Cuddapa District — Ganjam District — Godavery District — Kistna District — Kurnul District — Madras District—Madura District—Malabar District—Nellore District—Salem District—Tanjore District—Tinnevelly District — Trichinopoly District — Vizigapatam District. 2. *Mysore and Coorg.* 3. *Native States.* Bunganapilly—Cochin — Puducottah — Sundur —Travancore. 4. *French Possessions.* Chandanagore — Carical — Mahi — Pondicherry — Yanaon. 5. *Ceylon.*

Boundaries and Subdivisions.—This division includes the whole of the mainland of India south of the Nizam's dominions, and the Bombay Presidency, with a considerable extension along the coast of the Bay of Bengal, known as the Northern Circars, reaching the Chilka Lake, and shut in by the Central Provinces. Omitting this extension (which consists of a narrow strip of coast, called the Golconda and Orissa coast, whose width is nowhere more than eighty miles), it is a compact territory, of triangular shape, reaching from the 16th to the 8th parallel of latitude. It includes, also, the island of Ceylon. The total area is 204,162 square miles, and the population about thirty-two millions. Its extreme width towards the north is about 300 miles. It has a very extended coast line, partly on the west, partly on the south, and partly on the east coast, amounting in all to nearly 2,000 miles, exclusive of Ceylon. The number of good harbours is small, but there are no less than 144 places at which cargo is received and discharged.

Besides the mainland, this part of India includes the large and important island of Ceylon, which, however, has always been held by Great Britain as a colony, and not as a dependency. The whole of Southern India may be thus subdivided,—

	Area in sq. miles.
1. Madras Collectorates (British)	140,890
2. Mysore and Coorg	29,119
3. Native States	9,466
4. French Possessions	87
5. Ceylon (British colony)	24,600

General Account.—Southern India consists chiefly of a large tract of plateau, or table-land, greatly elevated on the western side, at the foot of the eastern escarpment of the Western Ghats, and also to the south, where the mountains cross the peninsula. This plateau slopes gradually to the east and north. In the southern and principal part it includes, also, the ridges of the Western Ghats, and their steep slope towards the Arabian Sea on the west, and throughout it includes the chain of the Eastern Ghats. The Neilgherry mountains, and the high lands in the southern extremity of the peninsula, render the whole broken and mountainous. It is crossed by many rivers, none of them very large, but several of considerable importance. The lower part of the course and the deltas of the Godavery and the Kistna, and many tributaries of the Kistna, the whole of the Cauvery and its tributaries, and the whole of the Pennar, the Palar, and many smaller streams, afford abundant opportunities for supplying water for irrigation, without which India would be a desert.

The climate of Southern India is greatly influenced by the comparative narrowness of the land, the mountainous character of the coasts, and the direction taken by the monsoon winds. This part of the peninsula is exposed to the full force both of the north-west and south-east monsoons; but much more rain falls on the western side, where the Ghats are highest, than on the east coast. Some districts, no doubt, such as Coimbatore and Mysore, share in the rain brought by both monsoons; but generally the interior of the country is dry, and those parts far removed from the sea, as Bellary, get exceedingly little rain. In point of temperature, the Madras Presidency is the hottest part of India. On the Malabar coast, where the atmosphere is moist, the temperature is seldom lower than 68°, or higher than 88°, the mean being 78°. On the Coromandel coast, however, the mean is 84°, rising occasionally to more than 100°. Previous to the rains the country is everywhere parched, and life is endured with difficulty. This happens in May, June, and July. In August the heat, dust, and glare are somewhat modified by occasional heavy thunder-storms. The great rains fall about the end of October, and then in a few days the surface of the whole country is changed, as if by magic, from a naked expanse to a sheet of the most varied and luxuriant verdure. Along the coast the sea breezes, which set in shortly after noon almost throughout the year, do much to moderate the temperature.

The people are of many races, but Hindu and Mahomedan races greatly prevail. By far the greater number profess the Hindu religion, and Bramins are more numerous than elsewhere in India. The languages spoken are Telugu, Tamil, Canarese, and Malayalma, besides Urdu and English, which are used everywhere. The education of the people is rapidly extending, the number of schools under inspection in 1866 having been nearly 1,400, and the number of pupils attending them more than 50,000. These are in addition to village schools under native management.

The internal communication by roads is very complete in the Madras Presidency. The roads are in excellent order, and at intervals of ten or twelve miles are places of shelter for travellers. The streams are crossed by bridges.

The rivers of Southern India are but little navigable, and afford no assistance to communication. The canals, though not on such a large scale as in the Bengal Presidency, and chiefly connected with irrigation works, include some connecting the lagoons on the coast, that assist in carrying on an extensive traffic in a very convenient manner. The extent of telegraph line is large and increasing. In 1866 it exceeded 1,500 miles.

The railway system of India is already well advanced in the South. From Madras, the Madras Railway is open towards the north-west as far as Bellary, and the branch of the Great Indian Peninsula line, now open from Bombay to Kulburga, in Hydrabad, will before long be completed to meet the Madras Railway. The main line of the Madras Railway crosses the peninsula towards the south-west to Beypur, near Calicut, on the Malabar coast, with branches to Bangalore (in Mysore) and Negapatam, on the Gulf of Manar, opposite Ceylon (*see* p. 40).

In no part of India is the system of irrigation more necessary than in the South, and, on the whole, it is effectually carried out by the agency of the rivers. The Godavery and the Kistna have been more especially employed for this purpose, by means of a system of dams and distributing canals. The works on the former river were commenced in 1847, and on the latter in 1851. They are now completed, and have not only secured the irrigation, but have rendered great assistance to the internal navigation of the country. In the year 1859, a guarantee of interest on the capital invested was given to a company for the purpose of constructing works of irrigation in the Madras Presidency.

The Presidency of Madras is divided into five parts. These are subdivided into twenty districts (including Madras city), under the regulation system, and two non-regulation districts (Ganjam and Vizagapatam).

MADRAS, the capital of the Presidency, and the seat of British government in Southern India, is a town of about 450,000 inhabitants, on the Coromandel coast, in lat. 13° 5′ N., long. 80° 16′ E. Its situation is naturally very disadvantageous for commerce, as during two months in the year (November and December) there is no communication possible between large ships and the shore. Ordinary European boats can never approach the shore with safety, and landing in the native craft through the heavy surf is never without risk, which is increased by the multitudes of sharks always ready to devour any unfortunate victims. There is now a well constructed pier, and the communication is much better; but ships are still obliged to anchor under very unfavourable circumstances. The first view of Madras is striking, as many of the public buildings extend in a line fronting the sea, immediately opposite the landing-place. Behind these buildings are trees, which relieve the white stucco with which the buildings are faced. The beach is studded with houses of business. On one side is the black town, partly concealed by plantations; on the other Fort St. George, having the appearance of a fort, but overtopped by many lofty structures, connected with business, that rise within it. The extreme length is nearly four miles, the average breadth two and a quarter miles; but a large portion of the area is occupied by gardens and enclosures, the buildings being rather thinly scattered. There is a cathedral and many English churches, a college, barrack, and hospitals. Madras is altogether a modern town, built on a piece of land obtained by the English in 1639. This ground was originally the site only of the fort, but the town has grown around it. The trade of Madras is considerable, and it now enjoys direct railway communication to various places in the interior of the peninsula and to the western coast. Steamers leave regularly for almost every principal port in India; and there is great trade with Britain, Calcutta, Burma, the Straits, America, and the Mauritius. The chief exports are cotton, indigo, oil-seeds, coffee, sugar, rice, and coloured handkerchiefs. Distance from Calcutta, 885 miles; from Bombay, 640 miles.

1. MADRAS COLLECTORATES.

Arcot, North (lat. 12° 22′—14° 11′; long. 78° 17′—80° 12′; area, 7,526 sq. m.; pop. 1,500,000).—A district forming part of the slope from the Eastern Ghats towards the sea. The western portion, reaching the watershed, is broken into isolated ranges of hills of moderate elevation, but towards the east it becomes low and flat. In the northern part of the district, the Naggery Hills are prominently seen from a distance at sea. Several streams cross it during the rainy season, but they are all dry the rest of the year. The soil on the plains is sandy, mixed with clay and gravel, and is extensively cultivated with rice and other grain crops. Even on the mountains there are fertile tracts. Reservoirs for keeping back the water are very numerous, and the dimensions of some are extreme. That of Cauvery pak is eight miles long and three broad, retaining the waters of the Cauvery. Some of these reservoirs feed irrigation canals, and others navigable canals. The town of *Arcot* is on the right side of the river Palar, the largest of those that cross the district, and on the line of railway from Madras to Beypur. It was formerly the capital of the Carnatic,* under

* The division of Southern India thus named is not very accurately defined. It is generally understood to mean the whole of the district on the east side of the Western Ghats, from about the sixteenth parallel of latitude southwards to Cape Comorin. It was the ancient Hindu kingdom "Carnata," in which the Canara language was spoken. The district is historic, as it was long the seat of the war for empire in the East between the English and the French. Up to the end of the last century there was a Nawab of the Carnatic, who enjoyed actual power. His successors held the title, without the power, till 1855, when the last of them died. The dynasty was Mahomedan. Some writers have recognized a division into Northern, Central, and Southern Carnatic.

Mahomedan rule. It is a large improving town, well supplied with water, and provided with an extensive barrack. There are the remains of the old fort of the Nawab's palace, and of some mosques. The fort was taken and held by Clive, in 1751, under circumstances of extraordinary bravery and talent. Distance W. from Madras, sixty-five miles. *Arni* is another town, with a British cantonment, celebrated in Indian history. It was formerly a strong fortress. *Chittore* is considered the capital, but is not remarkable. It is eighty miles W. of Madras, and 1,100 feet above the sea. *Tripetty*, fifty-one miles N. by E. of Arcot, has a remarkable Hindu temple. *Vellore* is an important town, with a strong fort on the right side of the river Palar. It has also a splendid pagoda, dedicated to Kistna. There is a population of about 50,000. The town is clean and airy, and has an extensive bazaar. The fort is very large. The place is hot, but healthy. Distance W. from Madras, seventy-nine miles.

Arcot, South (lat. 11° 11′—12° 39′; long. 78° 42′—80° 4′; area, 4,933 sq. m.; pop. above a million).—Like North Arcot, it consists of part of the slope from the Eastern Ghats to the Bay of Bengal. It is hilly towards the interior, but flat near the coast, and is crossed by several streams, all of which are dry in summer, except the Coleroon, the southern boundary, which is always abundantly supplied and made useful for irrigation. The climate is moderate and equable, and less subject than in other districts to the storms generally common on the Coromandel coast. In summer, however, the dryness of the air is excessive. The French settlement of *Pondicherry* (*see* p. 239) is within this collectorate. The chief place is *Cuddalore*, on the estuary of the South Pennar. It has a small harbour and pier. The town is low, but not unhealthy; there are broad regular streets and good houses. Distance S. from Madras, 100 miles. *Fort St. David*, once the capital of the British possessions on the Coromandel coast, is three miles south. It was formerly called *Tegnapatam*, and was destroyed by the French in 1758. *Gingi*, eighty-two miles SW. of Madras, is a hill-fortress of some strength.

Bellary (lat. 13° 40′—15° 58′; long. 75° 44′—78° 19′; area, 11,351 sq. m.; pop. 1,250,000).—A very large district in the interior of the Presidency, bounded on the north by the Nizam's territory, and on the south by Mysore. It is altogether a highland, the most elevated part being to the west, where it approaches or reaches the Western Ghats; and to the south, towards the Mysore plateau. It is crossed by several streams, all entering the Bay of Bengal. The climate is exceedingly dry—more so, in fact, than in any part of India. The soil is very fertile. A distinguishing feature of the district is the large number of dark-coloured granite rocks that start up abruptly from the ground in the most fantastic shapes. The town of *Bellary* (pop. 35,000) is the head-quarters of the ceded districts. It consists of a fortified rock, several forts, and the native town. It is 1,600 feet above the sea; distance NW. from Madras, 270 miles. *Adoni*, or *Adwanni*, near the north-western frontier, is one of the hill-forts, of considerable extent. *Gooty* is a cluster of fortified hills, nearly surrounding a native town. The town is nearly 1,000 feet above the plain, and more than 2,000 feet above the sea. The population is under 5,000. Bellary is one of the districts ceded by the Nizam in the year 1800.

Canara, South (lat. 12° 11′—13° 39′; long. 74° 45′—75° 42′; area, 3,480 sq. m.; pop. about 500,000).—A narrow strip of hilly and very fertile country, with more than 100 miles of sea-coast, and only one seaport (Mangalore), forming the northernmost district of the Madras Presidency, on the west or Malabar coast. The population is very varied, including Bramins, Nairs, Moplays, Jains, Corars, and Christians: Jains are especially numerous.

There are many small inlets on the coast, that afford shelter for small fishing-boats. *Mangalore* is the chief town. It is built on the north side of an estuary; but there is a sand-bank at the entrance, greatly detracting from the value of the place as a harbour. The town is large, but the houses are mean, and there are no public buildings of importance. Teak and sandal wood are shipped from Mangalore. The sandal wood is from the Mysore Hills. Near the town, to the north, is a valuable deposit of porcelain clay, of the finest kind. Population about 12,000. Distance SE. from Bombay, 440 miles; W. from Madras, 370 miles.

Coimbatore (lat, 10° 14′—12° 19′; long. 76° 36′—78° 16′; area, 8,099 sq. m.; pop. 1,250,000).—A district in the interior of Southern India, bounded on the north by Mysore, and on the south by Madura and Travancore. The general physical aspect is that of a great recess opening to the east. It is on the whole level, the elevation of the plain where it touches the Western Ghats being 850 feet. Through the gap or recess opening to the east runs the railway (Madras to Beypur), and also a stream carrying off the rains of the monsoons. All the hills, except where cultivated, are thickly covered with forest, abounding in teak and other valuable timber, and frequented by elephants, tigers, cheetahs, and other wild animals. In the western part of the district rise the Neilgherry Hills,* which, from their central position, their

* These mountains (*see* p. 8) form a triangular mountain mass, covering an area of about 600 square miles. The north side is connected with the table-land of Mysore, by a neck of high land, about fifteen miles wide. The mass is otherwise detached. From Coimbatore, the mountains rise in a vast precipitous mass to the height of 5,000 to 7,000 feet. There is no natural lake, but a large reservoir has been constructed by embankment in the vicinity of Ootacamund. The Neilgherries are

great elevation, and the evenness of their seasons, have become the resort of Europeans from all parts of India. Coffee, tea, cinchona, and many European fruits and vegetables are grown on the hills; while cotton, tobacco, grain, millet, and the castor-oil plant are cultivated in the plains. Iron ore, beryl, and saltpetre are obtained in large quantity. The only important manufacture is that of cotton cloths. The town of *Coimbatore* is at the foot of the Neilgherries, on the left bank of a small tributary of the Cauvery. It is well built, with clean wide streets, and well ventilated, but has indifferent water. It is 1,483 feet above the sea; but the railway from Madras to Beypur passes it. Distance SW. from Madras, 266 miles. There is a reservoir next it, forming a large lake during the rains. *Bowani-kudal*, at the confluence of the Bowani and Cauvery, has two very celebrated temples. *Darapuram* is a well-built town, with good streets, 42 miles SE. of Coimbatore.

Cuddapa (lat. 13° 12′—16° 19′; long. 77° 52′—79° 48′; area, 9,140 sq. m.; pop. 1,000,000).—A large district in the interior, adjoining Mysore, consisting of extensive plains, sloping from nearly 1,200 feet above the sea in the west, to above 500 feet in the east. The mountains of the Eastern Ghats rise in numerous parallel and continuous ridges abruptly from these plains. Numerous streams and watercourses cross the plains, and unite to form the river Pennar, a considerable stream, which breaks

---

not densely wooded, but are crowded with animal life. The human population, though scanty, includes five distinct races, *Erulars* and *Kurumbars*—savages like the Bheels; *Kotars*, a very peculiar race, exercising certain handicrafts, and not admitting caste; *Burghers*, the wealthiest and most civilized (they are Brahminists); and the *Todars*, or *Toruwars*, who hold sacred offices. There are six passes into the Neilgherries. Coffee has been very extensively cultivated in the Neilgherries for some years past, and is now an important crop.

through the Eastern Ghats, and enters the Bay of Bengal a little below Nellore. The soil is fertile, much of it consisting of the well-known *regur*, or black cotton ground. Besides cotton, rice, tobacco, oil-seeds, cardamoms, indigo, and sugar-cane are extensively cultivated. Mango, tamarind, plantain, and water-melon are among the most common fruits. Guava, peach, lime, citron. pomegranate, and grapes are less common. The chief manufactures are cotton goods, coarse woollens, the preparation of indigo, working in metals, and pottery. The soil contains much soda, salt, and saltpetre. The district is traversed by a railway, and there are good ordinary roads. The town of *Cuddapa* is built on a slope near the banks of the Pennar, and is well built and healthy. A large military force is stationed there, and the railway passes close to the town. Distance from Madras NW., 139 miles. *Rachuti*, on a tributary of the Pennar, 30 miles S. of Cuddapah, is another town. The district of Cuddapa is one of those ceded by the Nizam in the year 1800.

Ganjam* (lat. 18° 13′—19° 52′; long. 83° 50′—85° 15′; area, 7,657 sq. m.; pop. 1,000,000).—The most northerly district of the Madras Presidency. It is almost surrounded by the ancient territory of Orissa (*see* p. 49), and has a long line of coast commencing with part of the lagoon called *Chilka*,† where it touches the province of Cuttack, in the Presidency of Bengal. The coast is bold and rocky, dying away towards the north, where the sandy plain occupied by the lagoon commences. The district is the

---

* Ganjam and Vizagapatam are the non-regulation districts.

† The Chilka lake has been already alluded to in the description of lakes (*see* p. 20). It is saline, and everywhere very shallow, not more than six feet in the deepest part. It contains several inhabited islands, and is separated from the sea only by a narrow strip of sand. Excellent salt is made from it by evaporation. The Hindoo word for lake is *jhil*. Hence the name *Chilka*, or *jhil-ka*.

northernmost part of what is called the *Circars*.* The surface of the country is undulating, rising towards the west into hills covered with jungle, and containing the sources of various streams which water the plains. The level country is extremely fertile, yielding rice, sugar-cane, maize, millet, oil-seeds, cotton, &c., in great abundance. The hilly ground yields wax, lac, gums, dye-stuffs, arrowroot, and much timber and valuable ornamental wood. The inhabitants of the hills are Konds. *Chicacole* is the chief town, straggling and irregularly built with crooked narrow streets, overflowed in rainy weather. It is noted for its muslins. Population, 50,000. Distance SW. from Calcutta, 415 miles; NE. from Madras, 435 miles. *Berhampur* is a large town and cantonment, with a population of 20,000. The streets are narrow, dirty, and mean; but there are good bazaars. Silk and cotton cloths, sugar and sugar-candy, are manufactured on a large scale. The town is on a plain, surrounded by hills at a distance of a few miles. It has a tolerable climate. Distance SE. from Ganjam, twenty miles. *Calingapatam* and *Gopaulpur* are rising seaports. *Ganjam*, near the Bay of Bengal, was formerly remarkable for its fine buildings, but is now much decayed, the place having been abandoned in favour of Chicacole in consequence of a fever. It has some trade, but has not recovered its former condition. Distance NE. from Madras, 536 miles. *Gumsur* is forty-three miles NW. of Ganjam. Till taken possession of by the English, it was the scene of occasional human sacrifices among the Konds. Around it is an extensive forest, abounding in valuable timber. *Russelkonda* is a modern town, with cantonment, fifty miles NE. of Ganjam, at the foot of a low hill not far from the coast. Except during March, April, and May, the climate is pleasant.

---

* The districts of Ganjam, Godavery, Kistna, and Vizagapatam were formerly known as the Northern Circars. They were obtained by the French in 1753, seized by Clive in 1759, and formally ceded to the English, by the Emperor of Delhi, in 1765.

Godavery (lat. 16° 19′—17° 21′; long. 80° 56′—82° 20′; area, 7,533 sq. m.; pop. 6,000,000).—This district is low and flat near the coast, but hilly towards the north and north-east. It includes the delta of the Godavery, and some extent of country beyond. It is exceedingly fertile, yielding rice, maize, millet, sugar-cane, cotton, indigo, and tobacco, in large quantities, both for home use and exportation. The navigation along the coast is dangerous, owing to shoals and shifting sands, but there is one good port, *Coringa*, at the mouth of an estuary. It is a place of considerable trade, and well situated for building and repairing small vessels. The lake *Colair* (fresh-water) is partly in this district, and partly in Kistna. It is twenty-five miles long and ten miles across, and contains many islands. The Godavery has been adapted to irrigate the low lands near its mouth by the construction of a dam. There are many towns in this district. *Coringa*, on a branch of the Godavery, is a place of considerable trade, and is convenient for repairing small vessels. It is, however, liable to be inundated in the rise of the river after heavy gales. Distance NE. from Madras, 290 miles; SW. from Calcutta, 562 miles. *Ellore* is a well built populous town, on a stream which falls into the Colair lake a few miles below. It is forty miles N. of Masulipatam. It has carpet manufactures. *Madapollam* is famous for its cotton fabrics. *Nursipur* is at the mouth of the southernmost of the main branches of the Godavery, and is a rising town and port. *Rajamundry*, the capital, is on high ground on the left bank of the Godavery. The town consists of one principal street, about half a mile long, and a number of mean ill built cross streets. There is a fort a little outside the town. The population is large, amounting to 20,000. Distance NE. from Madras, 285 miles. *Yandon*, a French settlement, is in this district (*see* p. 239).

Kistna (lat. 15° 45′—17° 10′; long. 79° 15′—81° 40′; area, 8,353 sq. m.; pop. 1,000,000).—A district near the

Coromandel coast, including the delta of the river so called. Near the coast, and for some distance inland, the land is very low, and in some places below the level of the sea, but westwards it rises into hills of considerable elevation, attaining their greatest height near the town of Condapilly. The Kistna is rendered available for irrigation by a system of dams and canals. The vegetable productions are rice, cholam, oil-seeds, turmeric, betel, tobacco, and cotton. Salt is made along the coast; cotton goods are manufactured to some extent; and iron is worked in the hills. *Masulipatam*, or *Bunder*, the chief town, is built on the north side of a branch of the Kistna, on the Golconda coast. It is built on an extensive plain, marshy and unhealthy, and there is no good water. The houses are large, and the place is clean. Its manufactures are printed cotton goods and snuff. Population 28,000. Distance N. from Madras, 215 miles. The town has been rebuilt since 1864, when it was destroyed by a wave produced during a violent storm on the coast. *Guntur* is a large town. *Condapilly* and *Vinukonda* are hill-forts.

Kurnul (lat. 14° 55′—16° 15′; long. 77° 47′—79° 15′; area, 7,984 sq. m.; pop. 1,000,000).—A long strip of country, on the south or right bank of the Kistna, which separates it from Mysore. It is entirely inland, and is hilly almost throughout, the hills producing teak and other valuable woods. Owing to its position, it receives but little rain from either monsoon, and would suffer much from drought but for an extensive system of irrigation works in the neighbourhood of the chief town. The town of *Kurnul* has a population of 20,000. There are other small towns. Kurnul is one of the districts ceded by the Nizam in the year 1800.

Madras (lat. 12° 15′—13° 41′; long. 79° 35′—80° 20′; area, 3,010 sq. m.; pop. 500,000).—This district, with the exception of a few rocky hills, is everywhere flat, and the soil, when well watered, is fertile. There are few streams, the water supply for irrigation during dry weather being stored in reservoirs, of which there are great numbers, some of them being of large dimensions. Rice, gram, and other grains, sugar-cane, and betel, are grown to some extent.

Madras, the capital of the Presidency, has been already described (*see* p. 210), but the following towns are also within the district: *Chingleput*, a large town and fortress on the Palar, thirty-six miles SW. of Madras, in a valley, the upper part of which contains a reservoir two miles in length. The fort is large, and the town, which consists of one long street, not very interesting. It is not unhealthy. *Conjeveram* is a Bramin town, with some celebrated pagodas, much frequented by pilgrims. Distance SW. from Madras, forty-two miles. *Mahabalipuram*, a town on the coast, thirty-three miles S. of Madras, built in honour of the god Bala. Here are extraordinary rock temples, covered with sculptures. *Pulicat* is a town on an island in the extensive lagoon called the Lake of Pulicat. The lake is thirty-three miles long from north to south, and eleven miles wide, where widest. The town is twenty-two miles north of Madras. It is chiefly inhabited by fishermen, who belong to a mixed Hindu and Mahomedan race called *Lubbays*, met with all along the Coromandel coast. *Sadras*, a large, but poor town, once a Dutch settlement, on the Coromandel coast, 42 miles S. of Madras. Three or four miles inland are the Sadras Hills. *Punamali* is a station for British troops, 13 miles WSW. of Madras.

Madura (lat. 9° 5′—10° 54′; long. 77° 15′—79° 15′; area, 8,716 sq. m.; pop. 2,000,000).—A district in the south-eastern part of the peninsula, opposite Ceylon. Its coast forms part of Palk's Straits to the north, and the Gulf of Manar to the south, between which are the low lands and narrow channels that connect Ceylon with the mainland. The northern part of the district is rocky and hilly; the southern part, an extensive plain, without a hill or conspicuous eminence. The north-western part forms the Pulnai mountains, or Vurragerry hills, and are the resort of invalids. There are streams, but no important rivers, crossing the district. The principal is the Vyga, and most of the others are dry in the dry season. The climate of the hills is mild and genial, seldom below 50° or above 75°. In the plains there is great dryness and heat. Although there is rain with both monsoons, the country sometimes suffers from drought. The soil near the sea is sandy, but, in the interior, black and fertile, and well suited for cotton. Sugar-cane, betel-nut, and tobacco, are grown besides many fruits and esculent vegetables. The people are Hindus, and speak Tamil. There are good roads. *Madura* is the capital. It is a well built town, of great antiquity, and contains many remarkable pagodas. It was formerly the chief seat of learning of Southern India. Weaving, and working in brass, are extensively carried on. Distance SW. from Madras, 215 miles. *Dindigul* is a well built town, on the slope of a hill, with a fort. It is well supplied with water. *Ramnad* is a decent town and fort, near the coast, with a Protestant church, sixty miles SE. of Madura, where much coarse cloth for native wear is made. It is the nearest town to Ceylon, and the line of telegraph to Ceylon branches at this point from the main line.

Malabar (lat. 10° 15′—12° 18′; long. 75° 15′—76° 55′; area, 6,261 sq. m.; pop. 1,500,000).—A narrow strip of country, 140 miles long, and less than forty miles wide, between the Western Ghats and the sea, with several harbours for small craft along the low sandy shore. The Ghats rise from the plain to an elevation of from 5,000 to 6,000 feet, their sides being clothed with magnificent forests of teak, black wood, and cedar, the timber of which is floated down to the coast by the torrents that rush down the slopes during the rainy season. The passes across the Ghats are steep and difficult, but very picturesque. In the plains, the soil is amazingly fertile, and produces rice, cardamoms, coffee, and pepper in great abundance; the latter especially.

The climate of the coast is warm and equable, ranging from 68° to 88°. March, April, and May are the hot months, the monsoons beginning in June with storm. There is a great variety of races of men in this district. Among them are *Nairs*, who are numerous and influential, and were long the rulers. Among their peculiar customs is the total absence of the marriage tie. The *Tiars* are chiefly cultivators, and are much despised by the Nairs. Both these are Hindus. The *Moplays*, or *Mapilas*, are Mahomedans, the descendants of some of the very earliest converts. They are fanatical and troublesome. A large proportion of the inhabitants are Christians, who were at one time very numerous along the whole of the west coast of India. They are partly Syrian Christians, and partly Roman Catholics. The former have a very simple faith, and they appear to be reasonable and moderate. They attribute their origin to the preaching of St. Thomas the Apostle. Besides these religionists, there are also Jews of two denominations: the Black Jews, who are of dark colour, and of very ancient date, and the White Jews, who date from a much more recent period. The people generally are engaged in trade or agriculture, as there are no manufactures. The name "Malabar Coast," applied to the west coast of India generally, is supposed to be derived from the word "Malayalam," or "skirting the hills."

The subdivision of *Wynad*, in the east of Malabar, is an elevated plateau, 1,100 square miles in extent, rising abruptly from the west, but sloping towards the east. It is exceedingly picturesque and fertile; the hills are covered with forests of valuable wood, and every little valley produces abundant crops. The whole is well cultivated, and is a remarkable instance of the result of energy and industry in a district once avoided as deadly, and left to the elephant, the tiger and the monkey.

Beypur is a port, rising into importance, situated on an estuary about six miles south of Calicut. There is railway communication with Madras, and iron and coal exist in the adjacent hills. *Calicut* is a seaport town on the open beach in the southern part of Malabar, formerly a place of great importance. The haven is filled up up with drifted sand, and the palaces of the princes, who once lived here, are not even indicated by ruins. Vasco de Gama, the first European navigator who visited India, touched here in 1498. The town was taken by Tippoo Sultan in 1789, who butchered the inhabitants, and destroyed the place. *Cannanore* is a seaport of great antiquity, in a small bay fifty miles north of Calicut. There is anchorage, but the shelter is not complete. It is populous, but irregularly built, and has considerable trade in pepper, grain, timber, and cocoa-nuts. The soil and climate are especially favourable to the growth of the cocoanut-palm. It is the principal station in Malabar district. *Cochin** is a town about a mile long and half a mile wide, at the entrance of a large series of lagoons, which extend north and south along the coast for 200 miles, and are separated from the sea only by a narrow strip of land, varying in breadth from a quarter of a mile to a mile. This system of lagoons is called the *Backwater* of Cochin (*see* p. 234). The lagoons are connected with several rivers, but the number of open communications to the sea is very small, the principal one being that on which Cochin is built. Owing to its peculiar situation, it is well adapted to ship-building, and Cochin is, indeed, the only place along the west coast of India, south of Bombay, where large ships can be built. The approach to the harbour is impeded by a bar, but is practicable for vessels drawing fifteen feet of water. In-

---

* The town is not subject to the small independent state bearing the same name (*see* p. 234).

side there is a depth of thirty feet. The streets are good and well built. There is much trade. The population is large, and includes Europeans, Hindus, Moplays, Arabs, Persians, Portuguese, and Jews. Distance SW. from Madras, 350 miles; SE. from Bombay, 665 miles.

Mahe is a French possession (*see* p. 238). *Mananta-waddy*, in the Wynad, is beautifully situated, and is the resort of planters. It is forty-three miles NE. from Calicut, and fifty miles E. from Cannanore. *Tellicherry* is a seaport town, beautifully situated a little to the south of Cannanore, in a very healthy district. It is a busy place, and the neighbouring country is highly productive. The rainfall amounts to 130 inches during the wet season. *Palghat* is a straggling town, with a fort, in the beautiful valley bearing the same name, through which the railway runs.

Nellore (lat. 13° 55′—16°; long. 79° 8′—80° 21′; area, 8,341 sq. m.; pop. 1,000,000).—A flat district on the Coramandel coast, north of Northern Arcot, low and sandy near the coast, and comparatively barren, with large tracts of jungle, but rising in the interior into hills, few of which, however, exceed 400 feet above the sea. It is crossed by the Pennar river, whose bed is nearly dry in the dry season, and by a number of less important streams, all of which are dry, except during the monsoon, when the volume of water is extremely large. There are many reservoirs in the district, which assist in irrigation. The climate is dry, equable, and healthy; the fall of rain moderate (thirty or forty inches). Much of the land is uncultivated; but in the watered and cultivated parts all the ordinary crops are grown. There is salt made in large quantity, and iron and copper ores have been found. Among the inhabitants is a race called *Yanadis*, wild and savage, of short stature and black complexion, and living on roots and wild fruits, leaves, rats, and snakes. They are quite distinct from their neighbours in religion and language, and are probably aborigines. Nellore produces a fine breed of bullocks, much in request for draught all over Southern India.

Nellore, the chief town, is irregularly built; but it is a large town, with 20,000 inhabitants, and is tolerably clean. It is about 100 miles N. of Madras. *Ongole* is rather large, so far as population is concerned, but consists chiefly of miserable huts.

Salem (lat. 11° 2′—12° 54′; long. 77° 32′—79°; area, 7,610 sq. m.; pop. 1,200,000). A district in the interior, between South Arcot and Mysore. The western part is very mountainous, attaining in some places an elevation of between 5,000 and 6,000 feet above the sea; but there is much cultivation, and the forests are very valuable. In the eastern part, the line of the Eastern Ghats forms the boundary. There are some rivers and subordinate streams, and several reservoirs, in the district; most of the streams running southward and south-westward, into the Cauvery, but some northward, into the Palar. The hills are very healthy, and are visited by invalids; but the plains are swampy and unhealthy. The climate differs exceedingly in different parts. The chief products are cotton, coffee, tea, indigo, sugar, and tobacco. Carbonate of magnesia is found native in a stony barren plain in the middle of the district, and is extensively used for various purposes. Iron ore is also common in the hill country. The town of *Salem* lies in the narrowest and lowest part of a valley in the middle of the district. It is tolerably well built, with two wide handsome streets, and many of inferior construction. The population of the city is about 20,000. Distance SW. from Madras, 170 miles.

Tanjore (lat. 9° 52′—11° 53′; long. 78° 55′—79° 55 ; area, 3,720 sq. m.; pop. 2,000,000.)—A district occupying part of the south-eastern corner of the peninsula, on the northern side of the long line of rocky ledge that almost connects it with Ceylon. It has a sea-coast of 160 miles, including windings; and for half this distance the coast cannot be approached with safety, owing to the rocky character of the shore of Palk's Straits. In none of the ports is there shelter for other than small vessels. The tract is crossed by several streams, and is exceedingly fertile. There is much trade. Owing to an elaborate system of dams, cuts, and canals in connection with the Rivers Cauvery and Coleroon, there is always water for irrigation, and the soil is rendered exceedingly productive. *Tanjore*, built on an extensive plain on the banks of a branch of the Cauvery, consists of two forts, one of them four miles in circumference, with a fortified wall and ditch. The streets are irregularly built. There are numerous pagodas. The smaller fort is a mile in circuit, and contains within it a pagoda, considered to be the finest in India. It has considerable trade, and manufactures of silk, muslins, and cottons. Distance SW. from Madras, 180 miles. *Combaconum*, 20 miles NE. of Tanjore; large, and with many fine pagodas, and a provincial college. *Carical*, a French settlement (*see* p. 238). *Mayaveram*, a place of pilgrimage, with many pagodas, and a large population. *Najore* is on the coast. *Negapatam* is a place of considerable trade, on an estuary of the Cauvery, nearly opposite Ceylon. It is well built and improving, and is the starting point of a railway to Madras by Trichinopoly. *Sheally* is a large town, with many inhabitants. *Tranquebar* is a large town on the coast.

Tinnevelly (lat. 8° 9′—9° 56′; long. 77° 15′—78° 26′; area, 5,144 sq. m.; pop. 1,250,000).—The south-eastern extremity of the peninsula of India, with a coast line of about 100 miles, on the Gulf of Manar, and a country extending back to the ridge of the Western Ghats, at an elevation of 4,300 feet. Near the sea, the land is everywhere low and flat, but it gradually rises to a plain about 200 feet above the sea, and from this plain rise the mountains, from which many streams proceed, and water the district. During the monsoons much of the country is flooded, and a large part is watered by irrigation, the water being conveyed in canals from the River Tambaravari. The climate is very hot and dry, except at the season of the monsoons. The natural vegetation of the country includes enormous timber trees, date-palm, sago-palm, many twining plants, and others. The soil is poor. Cotton is grown, and the ordinary crops of the country are obtained, but the cultivation is neither very great nor very profitable. There is a pearl fishery on the coast, but that also is unimportant, and the navigation of the coast is too dangerous to admit of much trade. Among animals, the wild elephants are numerous, and sometimes troublesome. *Tinnevelly* is the principal town. It is situated in the interior of the country, on a stream; on the other side of which is *Palamcotta*, a military station. The population is 20,000. *Courtallum* is a small town in a recess on the eastern side of the Western Ghats, thirty-two miles from Tinnevelly, and 700 feet above the sea. It is the resort of invalids from the plains during the hot season. The scenery is very beautiful, and the beauty is enhanced by a number of waterfalls, a stream falling in a succession of leaps, the lowest of which is 200 feet down. *Tuticorin* is the only port of any importance. It has a safe roadstead, with good anchorage, sheltered on one side by the mainland, and on the other by a chain of islands, extending about eight miles from north to south. The shipping trade is considerable, cotton being the chief export. There are pearl banks adjacent, but they have not proved very important. Distance E. from Tinnevelly, 33 miles; SW. from Madras, 325 miles.

Trichinopoly (lat. 10° 37′—11° 31′; long. 78° 13′—79° 37′; area, 3,097 sq. m.; pop. about a million).—A small district on the eastern side of the peninsula, in the interior, and situated between Salem and Tanjore. It is flat, the flatness broken only by hummocks and rocky fragments of granite, projecting above the general surface. The granite is a useful stone, hard and very durable. Except in the extreme south of the district, the soil is a deep black fertile mould, producing two crops annually. The mean annual rain-fall is considerable, but not excessive, but the district is arid, and without irrigation would be a desert. For months together the temperature is always high, the sky cloudless, the air dry, close, and sultry, with much glare and intense radiation of heat. High winds and whirlwinds occur occasionally, but the monsoons are not strongly marked. Fogs and noxious exhalations are, however, unknown. The principal stream is the Cauvery, which is nearly dry in the dry season. Trichinopoly is subject to a plague of ants, and other destructive insects, and has several poisonous snakes. The crops grown are chiefly tobacco and cotton. Cocoanut-trees are very abundant. Rice, the grain called *raji* (*Eleusina coracana*), many kinds of millet, maize, and plaintains, are the food plants. The chief town is also *Trichinopoly*, situated on the Cauvery. The streets are straight and wide, but the houses low, small, and closely huddled together. Most of the streets have bazaars. Near the town, on a rock 600 feet above the plain, is the fort, including the citadel, a pagoda much resorted to by devotees, and other buildings. The inhabitants of the town manufacture filagree work, hardware, cutlery, saddlery, and cheroots. Close to the town, on an island formed by the embranchment of the Cauvery, is a celebrated pagoda, enclosed by seven square walls, the outermost four miles in circuit, and each twenty-five feet thick. Within the area, besides the pagoda, are streets and shops and residences of the Bramins. The town is called *Seringam*. Trichinopoly is 190 miles SW. of Madras.

Vizagapatam * (lat. 17° 15′—19° 3′; long. 82° 24′—84°; area, 18,935 sq. m.; pop. 1,250,000).—A belt of land on the coast of the Bay of Bengal, in the northern part

* Vizagapatam and Ganjam are the non-regulation districts.

of the Presidency of Madras, immediately south of the state of Ganjam. It is a part of the large extent of shore known as the Orissa coast. The coast line is bold, steep, and marked by a ridge of rocky hills. The surface of the district is generally undulating, rising towards the interior, and crossed by streams, which are dry except during the rainy season. The climate on the coast is hot, moist and relaxing; more inland it is equally sultry, though drier. The land-winds, however, which are generally oppressive in the Carnatic, are not felt. There are few industries. The crops are those common to the country. Iron is found in many parts, and cotton goods are manufactured. The town of *Vizagapatam*, the chief town of the district, is on the coast, and has a harbour and docks for small shipping. It is famed for its elegant elk-horn boxes and carving, and for its gold and silver filagree work. There is a fort containing barracks. The native town contains good streets, but is crowded. The climate is unhealthy. Near at hand is the civil station and cantonment of *Waltair*. Distance NE. from Madras, 380 miles; SW. from Calcutta, 470 miles. *Vizianagram* is thirty miles to the NE. It is healthy. There is a fort and military cantonment. *Bimlipatam* is a maritime town of rising importance, eighteen miles from Vizagapatam.

2. Mysore and Coorg.

These important principalities, which were till lately governed by separate princes or Rajas, are now under the control of a commission appointed by the Governor-General of India. They are extensive, and occupy a large part of the interior of the southern part of the peninsula. There are three principal divisions—Nundidroog, Astagram, and Nuggur. Each comprises two or three districts, and to the latter Coorg is added. Besides the Commissioner, there is a Judicial Commissioner, or Chief Judge, and a Director of Education.

Mysore (lat. 17°—15°; long. 74°—79°; area, 27,003 sq. m.; pop. 3,929,715).—An extensive country, in the interior of the peninsula, almost approaching the Malabar coast at its north-western extremity, bounded by the districts of the Bombay Presidency on the north, skirted by the Eastern and Western Ghats, and reaching the Neilgherry Hills on the south.

Mysore was erected into a state protected by the Indian Government in 1799, and remained for some time an independent principality, governed by a Maharaja. The late Maharaja, who died in 1868, assumed the government in 1810. In 1831 the Government of India marched troops into the province to subdue a formidable rebellion resulting from misrule, and the Raja was dethroned, and not subsequently restored. In 1865 he adopted a successor, who was recognized in 1867. This young prince is now about six years old, and is being trained under British influence for the government. It is at present under the protection and management of the British, and is managed by a Commissioner.

This country consists almost entirely of an undulating and elevated table-land, averaging 2,000 feet above the sea, and rising to nearly 6,000 feet in some places, the higher elevations consisting of isolated and detached ranges. It is very remarkable in its physical features, owing to the number of huge isolated granitic rocks, rising abruptly to heights varying from 1,000 to 1,500 feet above the plateau, with bases seldom exceeding two miles in circuit. These rocks are called *Droogs*, and are generally crowned with forts nearly inaccessible. Of these, some, as Nundidroog and Severndroog, exceed 4,000 feet in height above the sea. The mountain ranges are the Sivagunga (4,600 feet), and the Bababudin ranges (6,000). In the northern and north-western parts, the Ghats, receding, leave the intervening country plain and open, and sloping gradually, generally towards the north, but in one part westward. The drainage is to the north throughout the northern provinces, and in the southern it is to the south and south-east, the Cauvery being the principal stream, and its many affluents assisting to water the country.

The climate of Mysore is sensibly affected by its considerable elevation above the sea. At Bangalore (3,000 feet above the sea) the mean average at noon is 76°, and the maximum 82° in the shade; but the range is always great. The nights are seldom hot, the mornings and evenings always cool, if not cold, and the air is very elastic. The rain-fall is very heavy, the south-west monsoon bringing torrents, which fill the largest reservoirs in a few hours. Electric storms are common, and excessively violent.

The soil is a rich red earth containing iron in the heights, and a clayey earth in the valleys. It produces rice, ragi, millet, gram, wheat, sugar, betel, opium, and coffee. The irrigation is carried on by the aid of artificial reservoirs, which are very numerous. Some of them are surrounded by stone embankments. The water is good, when obtained from the river or reservoirs; but the well-water is brackish. In seasons of drought, as in 1866–7, there is famine and great suffering for want of sufficient storage of water. There are good roads through the country, and a branch of the Madras railway reaches Bangalore.

The people are a healthy robust race, taller than those of the Coromandel coast. They have regular features and fair complexions. Most of them are Braminists. They are generally regarded as deceitful, inconstant, and profligate; but they are courteous, contented, and patient under misfortune. They dress with a woollen blanket and a jacket, and they live in miserable mud-built houses, thatched, and without windows or other opening than the door.

Mysore, the chief town of the territory, is a decent, regularly-built, large town, on a declivity between two parallel ranges of hill. It is now only indifferently supplied with water, but has a canal leading from the Cauvery running through it. The wells yield impure and unwholesome water. There is a fort, within which was the Rajah's palace. The population is nearly 60,000. Elevation above the sea, 2,450 feet. Distance, ten miles SW. from Seringapatam. *Bangalore,* the chief station of the military force, is a well built town in the centre of Mysore, on a high ridge. It enjoys one of the finest climates in India, being cool and pleasant in the shade at all times. Distance NE. from Seringapatam, seventy-one miles. *Bednore,* or *Nuggur,* a considerable city, 4,000 feet above the sea-level; made the seat of Government by Hyder Ali in 1763, and from him called *Hyder-nuggur* (Hyder's town), since abbreviated to *Nuggur.* It is well situated for commerce, and was formerly so wealthy that Hyder Ali is said to have plundered it of property worth twelve millions sterling. Distance W. from Madras 360 miles. *Chittledroog,* a town and fort, the former on the plain, the latter on an adjacent rock of considerable size, part of a range of hills covered with jungle. The fort is strong, and long served as a state prison. It is 128 miles N. of Seringapatam. *Hunsur,* 13 miles SW. of Seringapatam, is a town noted for its manufacture of blankets and flannels. *Hurryhur,* a town and fort, once a place of considerable trade, and having a temple dedicated to the joint worship of Hari (Vishnu) and Hara (Siva). The climate is fairly good, and there is water. It is 132 miles NW. of Seringapatam. *Sera* is a town with a large, well built stone fort, within which are the remains of the residence of the former Nabob, and a large mosque. It has been the scene of fierce struggles. Distance N. from Seringapatam, 92 miles. *Seringapatam,* formerly the capital of Mysore, is situated on the extremity of an island on the river Cauvery, 2,412 feet above the sea. The town is ill built, mean, and dirty, badly ventilated, hot, with narrow streets, unhealthy, and excessively inconvenient. It was strongly fortified by Tippoo Sultan, whose palace was within the fort. Near it is the fine and lofty temple of Sri Ranga. Population about 12,000. The celebrated siege and taking of Seringapatam took place in 1799, after which Mysore fell into the hands of the English. Distance W. from Madras, 248 miles.

Coorg (lat. 11° 56′—12° 45′; long. 75° 25′—76° 13′; area, 2,116 sq. m.; pop. 120,000).—A rugged mountain district between Mysore and the Malabar coast, the lowest part of which is 3,000 feet above the sea. It is a constant succession of steep ridges and deep ravines, the whole clothed with forest more or less dense, but not so thick with underwood as to justify the term jungle, except to the east, towards Mysore. The general declivity of the country is towards the north-east, as indicated by the course of the Cauvery, the principal river, and its numerous feeders. The whole country is well intersected by roads, and is rapidly improving. The climate is generally healthy, and during part of the year remarkably equable, but is found to be unfavourable for the healing of flesh-wounds. It is wet, the rain-fall in 1835–6 being 119 inches. The crops include cardamoms (growing wild), coffee, tea, cinchona (for Peruvian bark), and cotton. The population are chiefly Nairs. The Coorgs are a handsome athletic race, brave, industrious, and intelligent; and superior in physical development to the natives of the plains.

Coorg is to all intents and purposes annexed to the British Empire, but is administered under the Commissioner of Mysore, for the benefit of the British Indian Government, as though it were still a native state under protection.

Merkera is the chief town of Coorg, 3,700 feet above the sea. It is situated on a table-land sloping gradually to the north and east, and dropping precipitously 500 or 600 feet to the lower country in other directions. On an isolated eminence is a fort. The climate is healthy, and the town is generally thickly peopled by visitors; the population being composed of the British authorities and their dependents. Distance W. from Madras, 315 miles. *Fraserpet*, a town on the Cauvery, 3,200 feet above the sea. The soil is alluvial and well drained, and the air salubrious, though warm in the day. The nights are always cool and pleasant. Distance W.

from Madras, 290 miles. *Virajenderpetta* a large town, 3,400 feet above the sea, chiefly inhabited by native Christians. It is sixteen miles S. of Merkera.

3. Native States.

The native states of Southern India are few in number, and, with the exception of Travancore and Cochin, very small in extent and importance.

Bunganapilly (lat. 15° 2′—15° 29′; long. 78° 8′—78° 27′; area, 500 sq. m.; pop. 35,200).—A small state in the district of Cuddapa, situated a little south of the Kistna.

Cochin (lat. 9° 48′—10° 50′; long. 76° 5′—76° 58′; area, 1,131 sq. m.; pop. 399,060).—A small irregularly-shaped hilly tract of country, between the British district of Malabar and the native state of Travancore, extending over the Western Ghats, and terminated westwards by a singular series of shallow lakes, or lagoons, called by the British residents *backwaters*, receiving the drainage of the numerous streams coming from the Ghats, but separated from the Arabian Sea by narrow spits of land. Owing to the torrents of rain that fall and run down towards the sea, during the wet season, these lagoons are liable to enormous changes of level and area. One of the feeders has been known to rise sixteen feet in twenty-four hours. These backwaters extend north and south for a distance of 120 miles, and by canals they admit of internal navigation to the extent of 200 miles from north to south. Their form is exceedingly irregular, and they communicate with the sea only at three points. They are always more or less navigable. All the lands around are occasionally swamped, and yield large crops of rice, or are covered with cocoanut-palms; but the atmosphere is rendered damp and unpleasant, though the coast is not considered particularly unhealthy.

The dense forests in the higher part of Cochin yield large quantities of valuable timber, which, with rice, pepper, and cardamoms, are the chief subjects of export. Coffee, cotton, and sugar-cane are also cultivated. The population is mixed, including Braminists of several castes, native Christians, Jews, and Mussulmans. There are also hill races, apparently descendants of the aborigines. All are poor and badly clothed. There are coarse manufactures of various kinds, and on the coast there are rope-makers and ship-builders. *Trichur*, the place of greatest importance, is well situated on the eastern coast of the backwater, and has several modern buildings. It is celebrated for its sanctity. Distance N. from Cochin, 41 miles.

Puducotta (lat. 10° 6′—10° 46′; long. 78° 33′—79° 16′; area, 1,037 sq. m.; pop. 62,000).—A small state to the north-east of Madura, subject to the supervision of the Madura Collector, but under the government of a Rajah, called the Tondiman. Much of it is covered with dense jungle. There is a town of the same name on the left bank of the Vellore river, 59 miles NE. by E. of Madura.

Sundur (area, 145 sq. m.; pop. 13,446).

Travancore (lat. 8° 4′—10° 21; long. 76° 14′—77° 38′; area, 6,653 sq. m.; pop. 1,250,000).—An important strip of land between the southern extremity of the Western Ghats and the sea. The Ghats rise in this extreme end of the peninsula of India a height of 7,000 feet above the sea, and terminate in a bold promontory a little above Cape Comorin. The principality of Travancore has a coast line of 155 miles to Cape Comorin, and throughout the whole distance there is no safe harbour for ships of any burthen, although there are several roads having good anchorage in favourable weather. On the coast there is a considerable extent of low country, in some places fifty miles wide, though generally less. It is much intersected by rocky hills, and has a mean elevation of 200 feet. Behind this, and towards the interior, there is a high plateau, 2,500 feet above the sea on its western edge, and rising gradually towards the east, where it reaches 4,000 or 5,000 feet.

Although Travancore is so near the equator, the high lands enjoy a moderate temperature, and even the lower parts of the country are cooled by the rains and by the proximity of the mountains and the sea. The thermometer does not appear often to rise above 90° in any season in the shade. The climate is moist, and the rainfall considerable; but though enervating, it is not absolutely unhealthy. There is, however, no bracing weather. The country is very rich in zoology, the variety of quadrupeds, birds, and reptiles in the forests being very great: tigers of enormous size, bats as large as chickens, the ant-bear, the black leopard, and many other rare and remarkable animals abound: snakes of the most deadly character, and alligators of great size are also common. The rivers and lakes, as well as the sea, abound with fish.

The soil in the low grounds is eminently favourable to vegetation. Rice and the sago-palm afford enormous supplies of food, and many vegetables are cultivated with success. The soil on the upper plateau is light and gravelly.

The population of Travancore includes Braminists, Mussulmans, and Christians, and a few Jews. The Bramins are very numerous, and include many of those called *Namburis*, considered to be aboriginal, and very highly regarded. There are also many Nairs, of the Sudra or labouring caste by descent, but engaged in various occupations, including the army. As is usual with these people, there are no marriage ties, the men and women living in promiscuous intercourse, without restraint. The Raja is said to belong to this class. The Mahomedans include Moplays, descendants of Arabs settled on the coast, Lubbis, mixed Arabs and Hindus, and a few of Patan descent. The Christians are of three classes—Ancient Syrian Church, Romanists, and Protestants, and are said to amount to an eighth of the whole population. Besides these, there is a wild race of savages living in the mountains.

Communication through this state is easy, as there are many good roads, several of them crossing the mountains to Madura and Tinnevelly, and not less than fourteen streams, all more or less navigable for country boats, crossing generally from east to west, and entering the lagoons or backwaters. The governing authority is a Raja; and Travancore is one of the most progressive and liberal of the countries in India still under native government. Police, hospitals, and schools have been long established; missionaries are freely admitted; restrictions on commerce are abolished, and there are many English settled in the country.

Trevandrum is the principal town. It is situated near the coast, not far from Cape Comorin, on the banks of a small stream. It is of considerable size, but is an ugly, ill built place. There is a fort about half a mile square, without a ditch, in which is the handsome palace of the Rajah, built in European style. An observatory was built in 1837, on a height adjacent the town. The town is 135 feet above the sea, and is 395 miles SW. of Madras.

Aleppi (*Aulapolay*) is the principal seaport. It is on the land between the lagoons and the sea, with which it communicates by a wide creek or inlet. It communicates with various towns in the interior by canals and the lagoons. It is a tolerably large place, without shelter for shipping, but having anchorage a few miles from the shore. Large quantities of pepper, cocoa-nuts, cardamoms, and timber are exported hence. Distance S. from Cochin city, 33 miles. *Quilon* is another seaport between Trevandum and Aleppi, where there is generally a small British force. It is a healthy town, with a population of 20,000, and much trade. It enjoys great facilities of water communication. It is said to have been the first settlement on the coast. *Travancore* was the former capital, but is now deserted. It is on the sea-coast.

4. French Possessions.

The French still retain in India a small number of towns and some surrounding territory, and these, with one exception, are on the eastern or Coromandel coast. With the same exception they are all within the Madras Presidency. The following are the various possessions:—

Carical.—A settlement within the limits of Tanjore, in the Presidency of Madras, situated near the Coromandel coast, on a small estuary of the Cauvery. The surrounding French territory contains an area of sixty-three square miles, and the population is about 50,000. The little port is quite shut up by a sand-bar during the dry season, but this is swept away by the first monsoon rains, causing an inundation of the river. It was restored in 1814, but is not allowed to be fortified.

Chandanagore.—A very small territory of 2,230 acres, delightfully situated on the banks of the Hoogly, in the Presidency of Bengal. The population is 35,000, of whom not much more than one per cent. are of European or mixed blood. It is subject to the Governor of Pondicherry. There is a French town, with an air of ruined greatness; its fine quay, and the wide street opening into it, being overgrown with grass, and the former residence of the Governor abandoned for one of less pretence. Adjoining the French town is a native town, consisting of huts and poor houses jumbled together with Braminical temples, having in front ghats, or flights of steps, giving access to the river, which is much frequented by bathers. Formerly, the Hoogly had sufficient water opposite the town to float ships of the line, but the deep channel has shifted. Chandanagore was occupied by the French in 1700, and flourished till captured by Clive in 1757. In 1763 it was given back, but retaken by the English in 1793, and finally restored in 1816. Distance N. from Calcutta, 17 miles.

Mahe.—A small settlement, with about two square miles of territory, within the limits of the British district of Malabar, on the Malabar coast, between Cannanore and Calicut. The population is about 3,000. It is on the south side of the estuary of a small stream only navigable for boats, but the site of the town is fine, being on high ground overlooking the river. It is neatly built. It was at one time troublesome, and was taken by the English in 1793, but given up again in 1815.

Pondicherry.—A territory 107 square miles in extent, on the Coromandel coast, eighty-six miles south of Madras, containing the town bearing the same name, and three districts, including in all nearly a hundred villages. The population in 1856 was about 120,000. The town of *Pondicherry*, the capital of the French possessions in India, and the seat of their supreme Government, is situated at the mouth of a small river, in a pleasant, convenient, healthy spot. It is a conspicuous object from the sea. There are two divisions of the town, one inhabited chiefly by Europeans, the other by natives. The former has well built streets and boulevards, and is regularly built. Pondicherry was first occupied by the French in 1672, having been obtained by purchase from the King of Bejapore. It was taken by the English in 1761, and restored finally in 1815.

Yanaon.—A settlement near the mouth of the Godavery, on the Orissa coast, in the Godavery district, within the Presidency of Madras. It is situated at the bifurcation of the navigable channel called the Coringa river, about nine miles above the embouchure of the main stream, and six miles from the town of Coringa. The extent of the territory is 8,147 acres (rather more than eleven square miles), more than half being land under cultivation, and 862 acres covered with woods and forests. Vessels of 200 tons can reach the town. The population in 1840 amounted to 6,881.

5. Ceylon.

This large and fertile island, lying at the entrance of the Bay of Bengal, and separated from the continent of India by the Gulf of Manar and Palk's Straits, lies between lat. 5° 56′—9° 46′ and long. 79° 36′—81° 58′. It is 270 miles in length, with an average breadth of 100 miles, being shaped something like a pear, narrowing towards the north. It is almost brought into communication with the mainland by a chain of sandbanks, connecting the two islands of Manar and Paumben, and called *Adam's Bridge*. The total area is 24,600 miles. It is divided into six provinces, and has a Governor, appointed by the Colonial Office, who is independent of the Viceroy of India.

The interior of Ceylon is mountainous, consisting of a lofty plateau, having a mean elevation of 6,000 feet, with various peaks rising to a further height of 2,000 feet. The highest point is Ellia, 8,260 feet, overlooking the high plain of Newera. Adam's Peak, further to the southward, is 7,420 feet. Among the mountains are beautiful and fertile valleys, and lofty plains. A broad belt of low land extends round the coast, and the northern half of the island is level. The northernmost extremity is broken into rocky islands, and terminates in the peninsula of Jaffna.

The largest river is the Mahavilli Gunga, which flows from the centre of the island towards the north-eastern coast, and has a length of about 200 miles. There are many smaller streams, and on the west coast are salt-water lagoons available for internal navigation.

The chief productions of the mineral kingdom are ores of iron and manganese. Plumbago, nitre, alum, and salt are also obtained. The island is rich in precious stones, yielding ruby and sapphire, diamond, cat's-eye, beryl, amethyst, garnet, and topaz. In the Gulf of Manar there is a valuable pearl fishery.

The climate is greatly influenced by the monsoons.

The north-east monsoon prevails from November to February, and the south-west from April to September. In the other months variable winds and calms prevail. The heat is, on the whole, less oppressive, and the temperature more equable than on the plains of India; but the eastern side of the island is hotter and more rainy than the western. In the mountains the climate is temperate.

The vegetation includes nearly all the valuable products of the mainland, besides others not there known, or flourishing less freely. Rice is grown in the lower parts of the island, and coffee in the interior provinces. The cocoanut-palm and the cinnamon-tree are characteristic. Cotton, the sugar-cane, and tobacco are grown everywhere.

Among animals the elephant is remarkable. The wild elephants are numerous and very large, and often destructive. All the animals common on the adjacent mainland, except the tiger, may be found in Ceylon.

The population is about 2,000,000. It consists of races of Hindu origin in the north and north-east. The Cingalese, or native Ceylonese, inhabit the central and southern part. Mahomedans of Arabic descent are found everywhere, and a savage aboriginal race called Veddas, live in the forests and remote parts of the interior. The Buddhist religion almost exclusively prevails, but Christianity has made some progress. The Tamil language is spoken in the north and north-east; Cingalese and Portuguese elsewhere. Agriculture is the chief occupation; but the Ceylonese have always been celebrated for gold and silver work, and for a kind of lacquered ware.

The revenue amounts to about £1,000,000 sterling, and is derived from land-tax, customs, stamps, salt, excise, and fisheries. The chief exports are cinnamon, coffee, and cocoanut-oil to Great Britain; betel-nut, timber, salt, and cocoanut fibre to other parts of India.

The means of communication are good, the island being crossed by excellent roads, and enjoying now also the benefit of a railway (*see* p. 40). Some of the roads pass along near the coast-line, others cross into the centre of the island. The scenery is generally charming, and sometimes magnificent. The road from Colombo and Kandy is especially fine as an engineering work, part of it running through a tunnel 500 feet long, cut through a mountain. A line of telegraph crosses the island from Galle to Colombo and Kandy, and thence to Manar, with a branch to the east coast. Crossing from Manar, the wire joins the main line through India, opposite Adam's Bridge, at the town of Ramnad, in the district of Madura.

Colombo is the seat of Government and the capital of the island. It is a neatly built flourishing town on the west coast in the southern part, and is defended by a strong fort. The harbour is only capable of receiving small vessels. *Caltura*, on the coast, a little south of Colombo, has some coasting trade. *Galle* is a seaport town situated on a low rocky point of land projecting into the sea, at the SW. extremity of the island. It has a spacious harbour, and is the regular calling-place for the steamers to and from Calcutta, Madras, Bombay, China, and Australia. *Jaffna* is a remarkably neat and clean town, on an island at the northern extremity of Ceylon. Much salt is made in the vicinity, and this and the timber of the black palmyra is largely exported to various parts of India. The port is called *Point Pedro*. *Kandy* is in the interior, near the centre of the island, seventy-two miles from Colombo. It is at the head of a valley, 1,700 feet above the sea, and surrounded by hills. It was formerly the capital. *Madura* is on the south coast, a little to the east of Galle. It has some coasting trade. *Negombo*, a little north of Colombo, is also a port with some trade. *Newera Ellia*, 6,200 feet above the sea, is the chief sanatorium of the island. *Trincomali*, on the northern part of the east coast, is small, but has a very fine harbour.

INDEX.

N.B.—The names printed in ITALICS *are not places. The first reference points in all cases to the description; the others are cross-references.*

It is suggested that if the name of a place be not found readily, the various possible ways of spelling according to the ordinary pronunciation should be tried, especially noticing that *i* = *ee* = *y*, *u* = *oo*, *c* = *k* = *s*; *h* and *r* are omitted where unnecessary, *o* and *a* sometimes interchanged.